PRAISE FOR *LIVING IN FLOW*

"Sky Nelson-Isaacs has created some excitement in *Living in Flow*, in which he explores meaning, perspective, authenticity, synchronicity, and all the things that make up the flow of our universe. He invites the reader in with a friendly, down-to-earth warmth, helping the nonscientist stay engaged. References to research and personal anecdotes make *Living in Flow* a compelling work that I enjoyed immensely."

—EDWARD VILJOEN, author of *Ordinary Goodness: The Surprisingly Effortless Path to Creating a Life of Meaning and Beauty*

"In this rare gift Sky Nelson-Isaacs has given us a remarkably insightful and readable understanding of synchronicity and how to live into it."

—LESLIE ALLAN COMBS, author of *Synchronicity: Through the Eyes of Science, Myth, and the Trickster*

"Reading this book made me happy. And hopeful. As an elder and a spiritual theologian it moves me to learn from a young physicist who is posing perennial questions from the viewpoint of the new science. Sky Nelson-Isaacs ignites new sparks that shed light on ancient mysteries. The author's method of LORRAX (Listen, Open, Reflect, Release, Act, XRepeat) as the path to flow parallels in many ways the four paths of creation spirituality in my work: Via Positiva, Via Negativa, Via Creativa, and Via Transformativa. There is rich and fertile ground here born of postmodern science for birthing a language that is far more friendly to spiritual and mystical realities than most of the perspectives of the modern world view ever was or could be. This book sparkles with insights and connections that we are all hungry for."

—MATTHEW FOX, author of *Sins of the Spirit, Blessings of the Flesh: Transforming Evil in Soul and Society*

"Sky Nelson-Isaacs brings a vigour and rigour to his analysis and locates his interpretation of synchronicity in day-to-day examples that have happened to him and to others, giving the reader a step-by-step analysis of synchronicity and how the theory applies. He develops an original model of synchronicity called 'meaningful history selection.' Grounded in quantum physics the visual model of the tree of possibility helps us understand how some choices are made en route, while others not, highlighting how we choose particular branches based on our intent."

—DR. PHILIP MERRY, author of the first grounded PhD research into synchronicity and leadership, and founder of Philip Merry Consulting Group, a Singapore global leadership consulting company

LIVING IN FLOW

The Science *of* Synchronicity *and* How Your Choices Shape Your World

SKY NELSON-ISAACS

North Atlantic Books
Berkeley, California

Copyright © 2019 by Sky Nelson-Isaacs. All rights reserved. No portion of this book, except for brief review, may be reproduced, stored in a retrieval system, or transmitted in any form or by any means—electronic, mechanical, photocopying, recording, or otherwise—without the written permission of the publisher. For information contact North Atlantic Books.

Published by
North Atlantic Books
Berkeley, California

Cover photo © gettyimages.com/MATJAZ SLANIC
Cover design by Howie Severson
Book design by Happenstance Type-O-Rama
Illustrations by Sky Nelson-Isaacs and Alison Manes

Printed in the United States of America

Living in Flow: The Science of Synchronicity and How Your Choices Shape Your World is sponsored and published by the Society for the Study of Native Arts and Sciences (dba North Atlantic Books), an educational nonprofit based in Berkeley, California, that collaborates with partners to develop cross-cultural perspectives, nurture holistic views of art, science, the humanities, and healing, and seed personal and global transformation by publishing work on the relationship of body, spirit, and nature.

North Atlantic Books' publications are available through most bookstores. For further information, visit our website at www.northatlanticbooks.com or call 800-733-3000.

Passage from *The Phantom Tollbooth* by Norton Juster text copyright © 1961, copyright renewed 1989 by Norton Juster. Used by permission of Random House Children's Books, a division of Penguin Random House LLC. All Rights Reserved. Reprinted by permission of Harper-Collins Publishers Ltd, © Norton Juster 1961; passages from *Synchronicity: The Inner Path of Leadership* copyright © 1996 by Joseph Jaworski, reprinted with permission of the publisher, Berrett-Koehler Publishers, Inc., San Francisco, CA. All rights reserved. www.bkconnection.com; Excerpt from pp. 85–86 from *The Bell Jar* by Sylvia Plath. Copyright © 1971 by Harper & Row, Publishers, Inc. Reprinted by permission of HarperCollins Publishers; Excerpt from *The Bell Jar* by Sylvia Plath reprinted by permission of Faber and Faber Ltd.; Excerpt of 8 l. from #48 from *Tao Te Ching by Lao Tzu, A New English Version, with Foreword and Notes*, by Stephen Mitchell. Translation copyright © 1988 by Stephen Mitchell. Reprinted by permission of HarperCollins Publishers; Excerpts from pp. 3, 6, 7, 44, 58–59 from *Flow: The Psychology of Optimal Experience* by Mihaly Csikszentmihalyi. Copyright © 1990 by Mihaly Csikszentmihalyi. Reprinted by permission of HarperCollins Publishers; Brief quote from p. 76 from *Creativity: Flow and the Psychology of Discovery and Invention* by Mihaly Csikszentmihalyi. Copyright © 1996 by Mihaly Csikszentmihalyi. Reprinted by permission of HarperCollins Publishers; Excerpts from pp. 69, 70, 119 from *Finding Flow: The Psychology of Engagement with Everyday Life* by Mihaly Csikszentmihalyi, copyright © 1997, 1998. Reprinted by permission of Basic Books, an imprint of Perseus Books, LLC, a subsidiary of Hachette Book Group, Inc.

Library of Congress Cataloging-in-Publication Data
Names: Nelson-Isaacs, Sky, author.
Title: Living in flow : the science of synchronicity and how your choices shape your world / Sky Nelson-Isaacs.
Description: Berkeley, California : North Atlantic Books, [2019] | Includes bibliographical references and index.
Identifiers: LCCN 2018034423 (print) | LCCN 2018036743 (ebook) | ISBN 9781623173128 (e-book) | ISBN 9781623173111 (pbk.)
Subjects: LCSH: Coincidence.
Classification: LCC BF1175 (ebook) | LCC BF1175 .N45 2019 (print) | DDC 123—dc23
LC record available at https://lccn.loc.gov/2018034423

1 2 3 4 5 6 7 8 9 KPC 23 22 21 20 19

Printed on recycled paper

North Atlantic Books is committed to the protection of our environment. We partner with FSC-certified printers using soy-based inks and print on recycled paper whenever possible.

*Dedicated to Rishi, Axel, and Sara,
who showed an early interest
in the ideas in this book.*

*To Dana, my soulmate.
I'm so glad I found you early.*

*To Ellie, for the future I want
you to have. You being here
makes it better already.*

CONTENTS

Foreword *ix*
Preface *xi*
Acknowledgments *xvii*

PART I

1. Searching for Meaning, Purpose, and Patterns *3*
2. Expect Synchronicity and Feel Flow *27*
3. Getting into Flow *61*
4. Building Symbolic Momentum *95*
5. Living from the Heart *133*
6. Authenticity as Flow *163*

PART II

7. Exploring the Foundations *189*
8. Meaningful History Selection *213*
9. You Are a Spark! *239*

A. Debunking Synchronicity *247*
B. Interpretations of Quantum Mechanics *257*
C. Calculating the Odds of a Synchronicity *263*

Glossary *267*
Notes *271*
Bibliography *277*
Index *285*
About the Author *299*

FOREWORD

I am deeply grateful that Sky Nelson-Isaacs has written this book. I have been waiting for over twenty years for a highly qualified physicist to embrace and verify the fact of synchronicity and to explain the fundamental physics that may underpin this phenomenon.

In these pages, Nelson-Isaacs has done so in a clear, concise, and nontechnical way—a way that all of us can understand. He has written with honesty, balance, passion, and a sense of urgency, given the state of the world we are now experiencing.

Most importantly, Nelson-Isaacs has given us a roadmap for cultivating the capacity to enable synchronicity in our everyday lives. The central message is that our choices shape our world. Every choice we make is meaningful because the Cosmos responds to the choices we make. Our intention and way of being—our total orientation of character—drive the future we will experience.

Accordingly, as we mature, we must answer the most important questions in life: Who am I? Why was I put on this earth? What is my life's purpose? Once we have determined the answers to these questions, we must live and act authentically, aligned with our purpose. The formula is: Listen carefully to what is seeking to emerge. Live from the heart—with love, wonder, boldness, and gratitude. And act—become a spark. When the inevitable challenges arise, have faith and remain committed. If we do this, we will experience success in uncommon terms.

These are life lessons that must be taught to our young people. They should be infused in every curriculum, beginning at an early age. Every organizational leader in business, in government, and in civil society must learn, embrace, and live by these principles. Today's national and global challenges demand this.

I congratulate Sky Nelson-Isaacs for his high service in writing this book. To the reader, I sincerely recommend this book and its profoundly important message.

—*Joseph Jaworski, author of* Synchronicity: The Inner Path of Leadership

PREFACE

Have you ever had a coincidental experience that made you say, "That was so weird"? If so, then I hope after you read this book you'll never feel compelled to say that again. Coincidences are only weird if they don't match our worldview, and I think the appropriate worldview for our age is one in which meaningful coincidence, or "synchronicity," is part of science. When synchronicity becomes part of our normal comings and goings, we find ourselves in that state social psychologists call "flow," a zone of "optimal experience."

The mainstream view is that these situations are just attributable to chance.[1] Some authors who have expanded on this view often look at a limited subset of dramatic examples, such as Carl Jung's scarab beetle,[2] sharing a birthday with someone at a party, or buying a red car and suddenly noticing red cars on the freeway. Then, under the assumption that these examples are sufficiently described by chance, the authors show that each single example is within reasonable odds. In some cases, known psychological phenomena such as cognitive bias already explain the situation.

But this approach leaves me feeling a bit confused when the alignment of circumstances feels especially meaningful to me. Having spent two decades seriously analyzing these experiences and considering all possible explanations (including self-deception), I have laid my bet on the possibility that the phenomenon is due to science, not statistics.

I have also read discussions of synchronicity that take the notion too far. They talk about the beneficence of the universe and the idea that we can "create our own reality." This proposition can be twisted into the notion that we are somehow fully responsible for everything that happens to us. If our life is hard, it must be our fault. Alternately, if I have my life together, I will have unending success. In my view, this view is short-sighted, reckless, and inaccurate.

My approach is different from the very start. I include a broad range of examples of synchronicity that are actually meaningful, in that they have an important impact on the people involved (unlike having the same birthday as someone at a party or seeing red cars after you buy one). I see synchronicity as a neutral process rather than a strictly positive one; it is a great teacher because it brings us circumstances that push us toward greater self-knowledge. It is not a fast track to success, but it *is* a fulfilling path to greater wholeness in our lives. I consider synchronicity to be a ubiquitous phenomenon rather than a quirk of nature, happening everywhere and part of everything. I am interested in everyday examples, not "weird" ones.

Each everyday example may reasonably be said to fall within the bounds of statistics, but when considered as a group, a pattern of bias toward meaningful situations seems to be a more reasonable explanation. I hope to show that these phenomena are consistent with current advances in physics, including research I have been a part of. My argument leads to a somewhat novel worldview strongly influenced by what we have learned from modern physics.

At the same time, the core ideas in this book may be familiar to traditions outside the Western view. Indigenous, Eastern and ancient mystical traditions are reflected to some degree in the conclusions I draw, and this may indicate that science is catching up to some of the understandings already illuminated by these ancient and worthy traditions.

Charles Eisenstein articulated today's dominant worldview thus: "You are a separate individual among other separate individuals in a universe that is separate from you as well. There is no purpose, only cause. The universe is at bottom blind and dead."[3] In other words, the world is a lifeless stage that doesn't know you are here. By contrast, the worldview I subscribe to holds that every choice you make is meaningful, each action builds momentum toward your conscious or unconscious goals, and these facts can have a real impact on day-to-day quality of life through the experience of flow.

Some of the ideas in this book are based on academic work in physical science. Those ideas are supported by papers on the foundations of quantum mechanics that have gone through peer review and can be considered more or less accepted by the scientific community. Other ideas reflect my own scientific research, which—although more recent and less well-established—has also been peer-reviewed and is fully consistent with the accepted body of knowledge.[4] The remainder of the book is based on my firsthand experience, interpreting what *I think* the science is telling us. Some of these pages step outside what I know with confidence and examine questions about what I think science tells us about human emotions and experience. Throughout this book, I encourage you to read both cautiously and openly. The field of quantum mechanics (and modern physics in general) is still very much alive and evolving, with fundamental questions that remain unanswered. This book should be read as my understanding of the currently known facts, my interpretation of their significance, and some of my original ideas for consideration.

I suspect that in writing this book I open myself to two major criticisms. Some may have already made up their minds about this subject and may feel that a study of synchronicity doesn't fall within the realm of physics. I will address this issue in the text to some extent. Ultimately, one should be careful about what one is sure of in science, because as scientists all we can cling to is the *method* of doing science, not the scientific conclusions themselves. The history of science is filled with examples of dead ends reached by well-meaning people who confused their models with reality itself and therefore couldn't see the way forward. In writing this book, it has not always been easy to find a path forward between the known and the unknown, but I have tried to do so with both respect and urgency.

The second potential criticism is that some may feel I am underqualified to do this research, as my educational path has been nontraditional for my field, physics. This is a valid point, as I do not have a PhD in physics or any other field. I can offer a twofold response, not as defense but as reassurance that I have taken some measure of care in coming to my conclusions. All my college degrees are in my field, and I earned them from respectable institutions (BA physics, University of California, Berkeley; single subject teaching credential, physics, Sonoma State University; MS physics, San Francisco State University). My career has involved many years of teaching physics and other science and math courses across a broad range of ages and skill levels. In addition, for the more than twenty years since I completed my undergraduate work, this field has remained my passion, and I have tried to stay current on its developments.

It is not easy to write a book that puts forth new ideas in science and simultaneously carries a message intended to be personally relevant to the average reader. I have organized the book in a way that attempts to separate the known from the unknown and the science from the personal message. Part I introduces the basic concepts of flow and synchronicity, and it discusses how I think these concepts can make a difference in the quality of our experience, from finding new opportunities for employment to living from the heart to finding our authentic selves. These chapters are overall more speculative, based on establishing novel relations among existing science, my new proposals, and personal experience. Part II provides a more in-depth exploration of the scientific research that I suggest supports these ideas. Some of this science is very well established, whereas other parts of it are more on the cutting edge and thus less well established.

I hope you, the reader, find in these pages the necessary balance of skepticism and boldness, head and heart, to want to examine the message for yourself and see if it fits. While the method of science is extremely powerful, it is not the only way of gaining understanding. I wish to approach this topic both with the tremendous clarity that physics can bring and with humility in the knowledge that we are not the first to tread on this path. Charles Eisenstein wrote eloquently about "the more beautiful world our hearts know is possible." For me, the message of the science of synchronicity and flow is about developing the motivation and willingness to try to create that world.

ACKNOWLEDGMENTS

I am grateful to Dana, my wife and "physician's assistant," for sticking with me and loving the real me, for her contributions to and feedback on these ideas, and for being open to hearing about it constantly for many, *many* years ("See that? *That's* synchronicity!"). Thanks to Ellie for her patience while I finished this project and for being a daily source of joy, inspiration, learning, and a reason to keep going. Thanks also to my extended family, especially my parents, stepparents and parents-in-law, and network of friends and fans.

Thanks to the Faggin Foundation for generously supporting my research over the past year. Thanks to Alison Knowles, Louis Swaim, Brent Winter, Bevin Donahue, and the team at North Atlantic Books for believing in the book, trusting me, and being excellent partners in delivering the finished work, and to Maurizio and Zaya Benazzo for the referral to North Atlantic. Thanks to Julie Barer for her professional support and guidance, Joseph Jaworski for writing the foreword, Alison Manes for creating the figures, Susan Pike for technical support, and all those who contributed their personal stories to the text. I'm especially grateful to Matt Upton, Daniel Sheehan, Menas Kafatos, Jeff Curtis, George Weissman, Brenda Dunne, and James Baldwin, who have been dedicated mentors to me.

Thanks to those who have edited my writing and offered invaluable feedback at various stages: Lizzie Moore, Laura Verrekia, Jai Flicker, Julia Mossbridge, Bernard Beitman, Ann Betz, Will Reid, David Strabala, Jude Rowe, Gaylen Moore,

Jeremy Lent, Jeremy Richardson, Gigi Azmy, Lianna Shannon, Rana Barar, Jon Zimmerman, Veronique Dellabruna, Samantha Estock, Gabriela Hofmeyer, Marius and Zsoka Scurtescu, Justin Riddle, Cynthia Sue Larson, Ann Marie Davis, Katie Dutcher, Bill Malady, Philip Merry, and Matthew Fox; the National Speakers Association community, especially Wendy Hanson, Janet Schieferdecker, Robin Weintraub, Alicia Berberich, Michael Lee, Rodney Dunican, and Kristi Matal; and my former students Joseph Dowd and Thomas Bischof. Thanks to Revs. Mary Murray Shelton, Gloria Conley, Barbara Leger, Tara Steele, Edward Viljoen, Karyl Huntley, the late Robin Gail, and many other community leaders who have provided me opportunities to refine my message.

Finally, thanks to Axel, Rishi, and Steve, my fathers, for always standing by me and helping me become a better dad myself.

PART I

SEARCHING FOR MEANING, PURPOSE, AND PATTERNS

Stephen Gaertner, a Czechoslovakian Jew living in Hamburg, Germany, was eight years old in 1937 when he came down with tuberculosis. Stephen's doctor advised him to go to a sanatorium in the Bavarian mountains, as was the common prescription of the day for treatment of tuberculosis. (Antibiotics were not yet fully developed; see the sidebar on p. 6.)

Even at that young age, Stephen had a sense of the unrest occurring in his country. He protested to his mother, "There would be *Hitlerjugend* [Hitler youth] in the sanatorium!" So his parents agreed to send him to Switzerland instead. A year later he was cured, and on March 9, 1938, his mother came to pick him up and take him back to Hamburg. But Germany had changed for the worse while Stephen was away. While his mother seemed to feel they were safe from the Nazi threat because they were not German citizens, he didn't want to return to Nazi Germany. He protested to his mother again, and she replied, "It's great winter weather, so I'll stay a week or two to ski; then we'll go." On March 15, news reached them that German troops had invaded Prague. Stephen's mother realized the danger and agreed to delay their departure indefinitely. They stayed in Switzerland until 1946, surviving the Holocaust unscathed. Stephen's father, who had remained in Hamburg, perished.

Stephen thinks about his experience in the following way: "Had the Nazis invaded a few days later, I would have gone back to Hamburg and perished with my father. That timing, together with my getting tuberculosis, saved my life."[5]

I propose that although such circumstances cannot be controlled or predicted, we can learn to navigate the flow of circumstances in a manner that defies chance by paying attention to synchronicity (or meaningful coincidence). This view is based on research (mine and others') in physics and is consistent with research in cognitive studies, psychology, and philosophy. It is far from proven or accepted in the scientific community, but I will try to demonstrate that these experiences are ubiquitous in everyday life and that the scientific view advanced here, which explains synchronicity on the basis of meaning, provides a better explanation than the mainstream worldview, which relies on chance accidents. Although the proposal may require adjusting ideas we take for granted both in science and in daily life, it does not conflict with any known theories or experimental data. Rather, it removes or clarifies certain aspects of things we assume to be true so we can understand what may be really going on beneath the hood.

Living in flow is a rich, complex process in which human values and experiences play an essential role. When has anyone ever been happy they got sick? Yet Stephen's illness was part of a chain of events in which he ultimately survived the Holocaust. I say that Stephen's "negative" experience contracting tuberculosis was just as much of a synchronicity as the "positive" experience of, say, bumping into a good friend at the airport or finding two quarters on the ground just in time to pay the parking meter.

A synchronicity is an event with significant consequences that is woven into our life in a meaningful way. "Meaningfulness" in this sense can be thought of as the degree to which an experience we have in the external world relates positively or negatively to a feeling or inner experience we have. We consider an event meaningful to us if it aligns or shares properties with values, needs, thoughts, feelings, emotions, or ideals that we have expressed recently or are on our minds. Often it can be difficult to embrace synchronicity because we get caught up in what the meaning of a situation *really* is. (I will define "meaning" in a more objective way in chapter 2.) The ultimate interpreter of meaning is our inner knowing, which comes from the thoughts in our head, the feelings in our heart, the sensation in our gut, and whatever other sources we have for making decisions.

The willingness of Stephen's mother to stay and ski for an extra two weeks, the timing of Hitler's invasion of Prague, and probably many other small twists can be seen as meaningful because they aligned with Stephen's gut sense that he was in danger. Together they can be seen as flow. Neither Stephen nor his mother knew what effect their decisions would have, but by attending carefully to the choices available to them in the moment, they avoided a threat to their lives.

What Is Flow, and Why Does It Matter?

The notion of flow was introduced to science through the work of Mihaly Csikszentmihalyi (if you don't speak Hungarian, the following approximate pronunciation may be helpful: "Me-high Cheek-sent-me-high").[6] Csikszentmihalyi defines flow as a human state of optimal functioning, a dynamic balance of challenge and skill. In the proper activities and under the proper conditions we become one with our lives, enacting "a complete focusing of attention on the task at hand—thus leaving no room in the mind for irrelevant information."[7] When we are in a state like this, thinking and feeling become integrated, with neither one controlling the other.

I think of flow in terms of the events or circumstances that happen in our lives. We can know we are in flow when events seem to happen in a meaningful way and the external aspects of life seem to fit together with the internal ones. Maybe an experience we want to have becomes possible due to some small opportunity that spontaneously shows up, or we suddenly recognize how the situation we are in serves our purpose. As a result, we naturally know what to do in each circumstance, not getting caught up in our choices.

As Joseph Jaworski explains it, when you decide on a direction for your life, "the people who come to you are the very people you need in relation to your commitment. Doors open, a sense of flow develops, and you find you are acting in a coherent field of people who may not even be aware of one another. You are not acting individually any longer, but out of the unfolding generative order."[8]

Don't Worry about Cleaning Up ...

When Alexander Fleming was researching the common *Staphylococcus* bacteria, he returned from vacation to find himself burdened with unusual cleanup duties in his lab due to the recent departure of his lab assistant. To his surprise, while cleaning Fleming found that a contaminated sample had grown a mold that had killed the bacteria in the dish. The mold was eventually identified as *Penicillium rubens*, which releases an antibiotic compound into its environment. This excretion, penicillin, altered the course of medical history as the first antibiotic treatment for scourges like meningitis, scarlet fever, and diphtheria.[9]

This sense of mutual relationship in flow between thinking and feeling extends to our surroundings as well. We enter into a dance with life—whether it is our tennis racket, our musical instrument, our teammates, or our family members—and find that the whole notion of control drops away. Instead of controlling our environment, we find ourselves in a symbiotic exchange, an act of mutual creation. Csikszentmihalyi says, paradoxically, "Thus the flow experience is typically described as involving a sense of control—or, more precisely, as lacking the sense of worry about losing control that is typical in many situations of normal life."[10] So flow is not about gaining control or surrendering control; it is about *transcending the sense of worry about control.*

However, to my knowledge Csikszentmihalyi doesn't mention the concept of meaningful coincidences or synchronicity as playing a role in the experience of flow. Carl Jung described synchronicity as the alignment between inner and outer experiences, a "falling together in time." More formally, a synchronicity consists of "events which are related to one another ... meaningfully, without there being any possibility of proving that this relation is a causal one."[11] In the definition I will use in this book, a synchronicity—or, equivalently, a "meaningful coincidence"—is an experience that was initially not very likely to occur but has become more probable because of its meaningful alignment with our personal (or collective) choices.

I view these two concepts, flow and synchronicity, as mutually dependent. In short, when we align with circumstance, circumstance aligns with us. Csikszentmihalyi's version of flow tells us how to align with circumstance by getting "into the zone," and Jung's version of synchronicity tells us how circumstances align with us when we do that. Together these concepts form my definition of flow.

Is flow about getting into alignment with life? Is it about facing challenges appropriate to our development? Is it about letting go of fears? Is it about treating each moment as precious? It is all of these things. Getting into a state of flow requires adaptability to life at any moment, and this requires inner work, such as openness and a healthy relationship with ourselves. When our mind is focused on some fear of the future, how are we to see the current circumstances clearly enough to notice a hidden opportunity? If we don't treat each moment as precious, how are we to see the forks in the road that happen at unexpected moments?

When we incorporate both flow and synchronicity into our way of life, we recover a sense of ease, connectedness, and joy even in the midst of high-stakes endeavors. These views are consistent with recent trends in organizational development.[12] Imagine if, at the same time that we strive to close a big deal, we can also be unattached to the outcome because we are confident we will get what we need from the deal. Our openness allows us to come to an agreement that all parties feel good about.

Underneath many of the problems facing us today lie personal choices—choices our ancestors have made to bring us here, and choices we make today. Bigger issues like traffic congestion, fossil-fuel dependency, food distribution, and energy efficiency are related to smaller decisions, such as where we prefer to work or shop, which career path we aim for, and where we send our kids to school or go on vacation. Many of us are not only dissatisfied with the immense global problems we face but also with the quality of our own lives.

I see a way to address these global challenges by connecting their solutions to smaller choices we already want to make in our personal lives. If we go for what we love in life, we bring a creative energy that has the potential to solve problems. If we go for what we love in life, we are more likely to be authentic, which empowers us to speak out for what is right and build healthy relationships. When I say *we*, I mean each of us reading this book. *We* are the hearts and souls of major corporations, small businesses, educational institutions, and countless other organizations that have the potential to do even more good in the world than they already do. When *we* are authentic, *we* are more likely to contribute openly to the "Pool of Shared Meaning"[13] and make space for others to do so as well. When *we* are authentic, *we* create change within our organizations from the inside, and we can make a broad impact on the world.

Why don't we live life to the fullest? Why don't we strive to have the career that calls to us? Why don't we take our relationships to deeper levels of authenticity? Certainly these are complex questions, but I want to point out one possible answer: we worry that it won't work out.

This is where synchronicity and flow come in. The way I see it, flow is about getting into alignment with our circumstances and understanding that the cosmos is, to an extent that I will carefully define, responding to our choices. Through a process I call "meaningful history selection," the events that come into our lives appear to be influenced by the very choices we make. I find that getting into flow allows me to trust that whatever path I choose, circumstances will arise to help me walk that path. This is not a whitewashing of difficulties but rather a willingness to step into difficulty and face all that life brings.

Many of the decisions we make in life revolve around the desire to feel safe and secure. From the level of national security all the way down to making enough money to send our children to day care, we need to feel a basic level of safety in order to be productive in life. To me, living in flow is a compelling way to tame the fear of the unknown and tango with the uncertainties of life. The more we can come to trust—not "the world" but *our dance with the world*—the more we can flow with the inevitable losses and disappointments that come while nourishing the constructive connections that are equally plentiful. This is not a naïve belief that the world is good but an empowered belief that we can aim for our highest vision and successfully navigate the territory we will have to cross.

On our journey, we are not left to our own devices. The central premise of this book is that being in flow leads to a greater chance of experiencing meaningful coincidences. These coincidences lead us further on the path of flow. From this view, the best way to remain safe is to learn to get into flow and dance with life. Positive results are not guaranteed. Bad things happen to good people every day, and nobody escapes mortality. Should we even want to? In the meantime, by living in flow we will live a more vibrant version of ourselves, transcending the need to control life and opening up to its bounty and the beauty of our own soul.

The Cosmos Is Responsive

What is the meaning of life? Do our actions have a purpose? Is the universe friendly?

Finding answers to these questions that work for everyone is probably impossible. I can't even be sure my own answers to these questions stay the same from day to day! I don't believe there is just one truth to be found. As a white person, as a male, as an American, as a Californian, I can only comprehend a certain slice of experiences with which I am familiar. My sense of what is meaningful or what constitutes a purposeful life probably varies from your sense of those things quite a bit, as do my tastes in clothes or music.

As a physicist, though, I am trained to look for patterns and for commonalities between things that seem completely distinct. While I can't say what is meaningful for others, I am curious to understand what makes something meaningful to anyone. Whether you are black, brown, white, indigenous, immigrant, female, male, younger, older, LGTBQ, cisgendered, Eastern, Western, Northern, Southern, or any other distinction that makes you unique, something drives you to make meaning of the events in your life.

Some of us are confident that the meaning we make is all inside our heads. Others are sure that meaning and guidance come from a deity. Many Western scientists conclude that the universe has no inherent purpose other than the gradual unfolding of events according to physical and statistical laws. The ancient yogic traditions—which should also be considered sciences, because they follow a rigorous process of repeatable experimentation on the inner states of human beings—see purpose all around us in the unfolding of karma.

I believe the third question above provides insight into the other two. Is the universe a friendly place? Amazingly, I believe this is a question that physics can tackle. The theory I will present, based on some well-established science as well as some new science and some speculative ideas, states that the universe is neither friendly, hostile, nor indifferent to us. Rather, it is *responsive*. We live in a cosmos that responds to our actions by bringing us more of the same. To oversimplify for a moment, if we act friendly to the world, we find that circumstances emerge that reinforce our belief that the world is friendly. Similarly, if we act hostile to the world, we find our perspective justified because events arise that confirm our preconceived notions. When we align with circumstances, circumstances align with us. We can call this flow.

> ### Are You a Dodge Mechanic?
>
> "My young daughter and I were crossing the desert in an old Dodge Explorer. We arrived at a campground at night, and when we arose in the morning, the campground was mostly empty. Then we found our van was making strange noises and wouldn't start. What were we going to do? As we deliberated, a disheveled man climbed out of the creek bed nearby and said, 'I could hear you're having trouble. I'm a Dodge mechanic; can I help?' He worked for a few minutes, patched the problem, and gave us instructions to take to the next service station for a permanent fix. Thank goodness just the right person was there when we needed him." (Story contributed by Anne Cummings Jacopetti)

However, it is not as simple as it initially sounds. The proposed process of meaningful history selection indicates that life will be punctuated by meaningful events. Even if we approach people with kindness, not every person we talk to will be kind in response. This is evident within the first five minutes of my day when I gently wake my eight-year-old daughter. I have a fifty-fifty chance of being snapped at, no matter how kind I am! Rather, life is a stream of events *accented* by useful growth opportunities. These growth opportunities pop up throughout our days, and it is these "singular events" that I am trying to understand in my research.

Singular events are like forks in the road. They are points at which our choice of action makes a significant difference in the course of future events. If we envision all possible outcomes existing on a tree of forked branches, singular events are the forking points where one major path diverges from another.

My view is that we can find the essence of any spiritual teaching through the simple practice of responding to the experiences of life and seeing how life responds to us. Our job is to see what meaningful lesson each experience provides. The most important question to me is this: How does the cosmos respond to what I choose to do? If we understand how the responsiveness of the cosmos works, then we can develop a more compelling relationship to meaning-making and purposefulness in our lives. Such a relationship will allow us to align ourselves more effectively within our personal relationships, our professional relationships, and our political relationships. In this way we can *live in flow.*

If the universe is indeed responsive to our actions, then we are the source of the meaning in our lives. Experiences outside serve as a mirror for experiences inside, and every event can be a meaningful opportunity for learning about ourselves.

I used to see this as a nice philosophy, but as a physicist I undertook a quest to understand how it can be true. The way I see it, every action is meaningful with respect to something, because every action leads to some outcomes and not to others. Doing the dishes and folding the laundry are meaningful with respect to making my home a comfortable place, because these activities logically lead toward futures in which my home is a place I'd be proud to have my parents visit, and they simultaneously lead away from futures where I am frustrated by tripping over my piles of dirty clothes. But these activities don't influence whether I get new clients, so they are not meaningful with respect to my professional goals.

In contrast, time spent on professional activities creates a clear difference between futures where my profession thrives and those in which it languishes, but it doesn't differentiate futures in which my house is in good repair from those in which it is derelict. Whether I get a promotion or not, that bedroom door latch still doesn't close properly. Every action is meaningful, but I must look carefully inside myself to know which specific meaning my actions exhibit and whether it is the meaning I intend to act upon.

Seeing action in this light, might there be some alignment between intentional prayer, karma, and the responsive cosmos? Through our actions we choose where to invest our effort, and through the proposed process of meaningful history selection we are more likely to experience situations that reflect those choices. It is *I* who bring meaning into the world by tapping into the deep reservoir of divine urges and translating my inner experiences to life in the world.

Rather than wondering what the meaning of my whole life is, I can wonder how to bring more meaning to everything I do. My current view of the meaning of life is to live purposefully. What a challenge this is! Living purposefully can happen anytime and anywhere because the responsive cosmos reinforces whatever choices I make. So in order to understand the meaning of life (in the working definition I provide here), I think we need to understand what it means to live purposefully.

Acting with a Sense of Purpose

Are you motivated to be purposeful in your life? If you are, you are not alone. There is a growing cultural alignment between who we are at home and who we are at work. Kathy Caprino, a business coach for professional women, surveyed her online community[14] and found that the top qualities these women aspire to are as much personal as they are professional. Caprino's respondents said their top ten values were happiness, money, freedom, inner peace, joy, balance, fulfillment, confidence, stability, and passion. These are not the values we would typically expect people to have if all they cared about was their workplace; they describe *who* the respondents want to *be*. Upon reading more deeply, eight of these ten values (everything but money and stability) relate to having a deeper sense of purpose in life. If you are just as interested in thriving purposefully as you are in making ends meet, you are in good company.

So what does it mean to act purposefully? Every action is meaningful in some way, but a *purposeful* action is more difficult to achieve. A purposeful action is one whose meaning is aligned with a coherent plan on your part. To act purposefully, you must know what you intend and then take actions in alignment with your intent. This alignment can be quite elusive to achieve. I suspect we have all had experiences where we gave someone advice that was intended to help them, but they interpreted our advice as criticism. Imagine me saying to my daughter, Ellie, "You'll be happier if you don't wear those mismatched leg warmers to school." My action is *meaningful* in that it distinguishes between futures where she wears the leg warmers and ones in which she doesn't. As a result of my simply uttering those words, I am bound to alter those probabilities one way or the other, whether she complies or rebels.

The problem I want to shed light on is that my action was not aligned with my intent, so the meaning of my comment is not what I intended. I want her to change her leg warmers because I am nervous she will be teased about wearing mismatched clothes. Because I had a hidden motive, my actions will probably result in her feeling more self-conscious than she otherwise would have. In fact, I have distinguished between futures where she is self-conscious and those in which she is carefree, which is not what I intended. The meaning of my actions is not in alignment with my conscious intent, so this is not a very purposeful remark on my part.

If my core intention is to build my daughter's self-confidence, then at each moment I can try to align my choice of words with that intent. I might instead say, "I notice you chose some very colorful leg warmers this morning, and they don't even match!" For my eight-year-old, being noticed and reflected—without being judged—reinforces her ability to make choices and builds her awareness of who she is. This time I have acted with a sense of purpose by aligning the eventual outcomes of my choices with the experiences I want her to have.

> ### Pay Attention to the Little People
>
> One time I went shopping for hardware with my daughter, seeking to replace a wasteful sprinkler head with a cap. I searched for a cap that would fit but couldn't find one. Ellie was only four, but I was stuck, so I told her what I was looking for. A moment later she handed me a small package she had found sitting in the wrong place on a shelf. It wasn't at all what I needed, so I tried to find the right place to put it back ... but wait; maybe it was something I could use after all? It turned out that Ellie had discovered a different solution that I hadn't considered, and it was even the right size fitting. Problem solved!

There is a difference between "finding our purpose" and "finding a sense of purpose." The first phrase implies a very big statement about life goals, as if there is something we are supposed to be doing with our lives, and we have to find the right thing. I like the second phrase better. I think the universe responds to the choices we make by bringing new events into our lives that match those choices. Therefore, each of our daily actions becomes really important. Making each action purposeful is a *habit we can develop*. It's like tending a garden. You aren't a factory farm, pumping out only one cash crop. You are a community garden, planting different seeds at different times and for different purposes. Some of your actions purposefully build love within your family; others of your actions invest in your success in your career. Still other corners of the garden hold seeds you plant purely for the growing of your own joy and fulfillment in life. Finding a sense of purpose focuses more on the "sense of" part than on the "purpose" part; the purpose can be any goal or intention you take on, but your sense of purpose is unique to *you*. Finding a sense of purpose is about finding the essence of yourself in everything you do.

The role of physics in this discussion is to present us with an understanding of the pliable nature of meaning in our actions. I will show that every day our choices select certain future outcomes that align with them. Meaningful events that we call synchronicities occur along the way, becoming more likely to happen due to the simple fact that they lead to those very future outcomes we have aligned ourselves with. As this book helps you develop a left-brained, rational understanding of how science may predict meaningful coincidences in your life, I hope you find inspiration to actively investigate what living purposefully means for you, in a cosmos that responds to your choices.

What Is a Synchronicity?

A synchronicity, in my definition of the word, is an event that seems meaningfully related to what is going on in our lives, especially related to the choices we have made or are trying to make. The synchronicity event seems unlikely to happen—it might make us ask, "What are the chances?"—but it becomes more likely, as we will see, because of its relationship with potential future experiences that are meaningful to us. In this way, synchronicities shape our world. Let me illustrate with an example.

In my physics masters program, I had a friend named Evita. We had talked about my research on meaningful coincidences a few times, but she was skeptical, as any physicist should be. During our last semester, she was waiting to hear back from the PhD programs she had applied to. Soon Evita learned she had been accepted at a great school, but it was in a very expensive town and was a long drive from her family. The school that she really wanted to attend the most, on the other hand, had put her on a wait list. This was an urgent concern, because the school she wanted most was near her home and would allow her to survive on a much smaller budget and be close to her family.

I wanted to tell Evita that I suspected if she put effort into following up with the school, a coincidence might unfold in her favor. But her skepticism of that concept made me restrain myself. Instead I just commiserated with her.

A week later Evita sent me a text saying it looked like she wouldn't get into the program she really wanted. At this point I felt remiss and gave her the advice I generally offer about finding more synchronicity in life. I said, "Act like you really, really want it, even if you think it can have no effect. Maybe drive to campus and seek out the right person to talk to. If you can't find the right person, just talk to anyone you meet there, and see what connections you make. Read papers by professors on campus you might want to do research with. Take initiative and be directed, but remain open-minded." My advice was based on the principle that intentional, directed action can facilitate the appearance of unexpected coincidences that support our goals.

Evita appreciated my attempt to help, but it seemed obvious that she didn't believe any of that would work. After all, the university she was applying to was an institution with clear policies and procedures. It was her prior preparation, not her attitude after applying, that would get her in.

At the end of the week Evita showed up to class with a big smile on her face: she had gotten in at the last minute! She told me she had indeed called the school, and amazingly, it turned out that the acting head of the physics department had gone to school with her mother's thesis advisor. It was a natural fit.

This is a synchronicity. It is a coincidence that is quite unpredictable but that solved her problem in a meaningful way. More generally, synchronicities can happen at all scales and sizes, from the incredibly meaningful to the barely noticeable. The key element is an experience that is not causal (i.e., the fact that Evita called the school did not *cause* the department head to be Evita's mother's thesis advisor) but that is meaningfully related (i.e., this set of facts solved the problem Evita was facing). There can be *indirect* causal links, such as the fact that this school was located in the city where Evita grew up, so that the link between her mother and the school is less of a surprise. Nonetheless, there is no direct causal link between Evita's call to the school and the identity of the current department head. The odds against this coincidence don't have to be astronomical for it to be considered a synchronicity. It is the common occurrence of minor synchronicities rather than the rare occurrence of major synchronicities that I think is most applicable to our lives, for this is the type of synchronicity we can incorporate into everyday choices.[15]

Evita thanked me for the advice, and I said "Thanks to synchronicity!" She replied, "Normally I would disagree with you, but in this case ..." This was a meaningful coincidence that will likely make a very big difference in Evita's life over the course of her PhD program and afterward. Instead of going to school far from home in an expensive living situation, she is close to home and more financially secure.

Here we encounter an inconsistency that will be present throughout this book. I claimed above, with no uncertainty, that Evita's experience was a synchronicity, because it was such a strange yet beneficial circumstance that arose for her. But the ultimate decision about whether a coincidence should be considered meaningful—and acted upon or not—comes from her *inner knowing*. In the theory of meaningful history selection, although meaningful circumstances are defined objectively rather than subjectively, that doesn't give one person permission to tell another person which events in their lives are meaningful and should be paid attention to. Ultimately, synchronicity is defined from the perspective of each individual, and synchronicity respects the autonomy of the individual. It is the individual's inner knowing that makes the final judgment.

There is a growing body of research associated with the phenomenon of synchronicity, from authors including Carl Jung,[16] F. David Peat,[17] Joseph Jaworski,[18] Allan Combs and Mark Holland,[19] Philip Merry,[20] Walter Baets,[21] Bernard Beitman,[22] Helene Shulman Lorenz,[23] W. Brian Arthur and colleagues,[24] Kirby Surprise,[25] Deepak Chopra,[26] Arthur Koestler,[27] and Paul Kammerer.[28] The existing approaches to understanding or applying synchronicity range across psychology, physics, statistics, parapsychology, psychiatry, mythology, and business.

If a synchronicity is seen as a single stand-alone event, we might write it off as a fluke. But when we continually pay attention to the events of our lives as if they carry useful information for us—which means observing what I call the responsiveness of the cosmos—we enter flow. Again, the central premise of this book is that flow and synchronicity are intimately related. Synchronicities are moments when circumstances align for us in a meaningful way, and a sense of flow emerges when we can integrate circumstances like these into our daily activities and decisions.

Existing Research on Flow States

You have probably had the experience of flow many times in your life—perhaps even many times each day. For me it often happens more easily late at night, when my planning brain slows down and my creative brain gets absorbed in reading a book or writing an article, working on a song or playing piano. When flow sets in, instead of the chatter in my head about priorities, all I hear is my own curiosity about learning and creating. I am absorbed in what I am doing, and I move from task to task without questioning what I should or shouldn't do. Even with a watertight task list created with itemized pros and cons designed to maximize my success, it's often the unexpected things I do when I am in flow that propel me forward into new, more interesting directions.

In these flow experiences, time passes differently. I often stay up long after midnight because I am absorbed in whatever project I have found. I don't notice the time passing. I don't have an agenda. I simply sit down with time and space around me, and a task emerges that draws me in. Instead of worrying about whether I am working on the most important thing in my life, I find that natural curiosity draws my attention toward whatever project I land on. I bring meaning to the thing I am doing, and my life feels purposeful. In flow, I naturally and easefully determine what task I should tackle next. I am absorbed in a task until I naturally sense that it is time for something else. In the flow, each thing has its timing. To try to do a project before it is ready to be started, or to push it further than it is ready to go, is to miss the flow.

> ### Gee, I Wish We Could Get That Money Back
>
> "My spouse and I had long fantasized about owning a country house. One day while driving through the countryside we came upon a lot that was perfect. The price was the exact amount we had just sold our Friesian stallion for, but we had used the money to pay down our current home's mortgage. We thought, 'Gee, we wish we hadn't paid down the mortgage,' and we began scheming how we could raise the money. But then we received a letter from the mortgage company along with our uncashed check, indicating that they didn't know what we wanted them to do with it. Needless to say, we tore up the check and bought the lot." (Story contributed by Betty)

Flow has a significant amount of research behind it in the fields of psychology, neurochemistry, and (I suggest) physics. In Csikszentmihalyi's seminal book on flow states, he says that while one is in the state of "complete focusing" referred to earlier, "one is able to forget all the unpleasant aspects of life."[29] Flow is intimately tied to the focus of our attention. Csikszentmihalyi continues: "Because most jobs, and home life in general, lack the pressing demands of flow experiences, concentration is rarely so intense that preoccupations and anxieties can be automatically ruled out." We get into flow states by focusing intently on the task at hand, letting ourselves be unaware of the consequences of failure or the pressures of tomorrow. This allows the concern about doing the job just right to fall away from our minds. Csikszentmihalyi describes his research participants' perspective on flow as a "state of mind when consciousness is harmoniously ordered, and they want to pursue whatever they are doing for its own sake."[30]

Inspired by Csikszentmihalyi's work, other researchers and authors have focused on the changes in our brain states that lead to greater focus and decreased inhibitions. The main idea animating this research is that optimal actions originate from a heightened mental or emotional state. Steven Kotler has written bestselling books on the role of flow states in developing human potential, especially in extreme sports.[31] The influence of our surroundings on our ability to achieve these highly creative states has also been studied.[32] Limb and Braun at Johns Hopkins University found that when jazz musicians improvise, various functions of the prefrontal cortex related to self-censure are inhibited.[33] Musicians know well that when they're improvising, trying to decide what to do next is the kiss of death. A performer gets into a state of flow to let their authentic expression come through.

The research on the psychological components of flow in artists, athletes, and other high-intensity performers is compelling. But this research typically treats the external world as a static canvas, as if the experience of flow was only a way to perceive the world differently. Such a perspective ignores the possibility that the environment is responsive to our actions. In a responsive cosmos, when we enter flow *different circumstances occur* than would have occurred otherwise. The choices we make are reflected in the external situations that appear. Flow is not only a matter of our interpretation of life (i.e., a positive outlook) but a state of being that can *influence events outside of ourselves.* A flow state emerges when we align with life, and then we find that life aligns with us too.

Getting into flow by giving our focused attention to the task at hand is not always easy to do. Csikszentmihalyi identified the challenge level of a task and the skill level of the performer as important parameters affecting our ability to find flow while performing that task.³⁴ If skill outweighs challenge, we become bored; if challenge outweighs skill, we become anxious. In addition, both skill and challenge levels must be high. Finding flow is about putting ourselves in situations that are appropriately challenging and that we are skilled enough to perform. Csikszentmihalyi calls this appropriate balance between skill level and challenge level the "flow channel."

Much of the time the tasks we face in life may feel repetitive and predictable, such that we can slide by with hardly a thought. Going to work on the subway is routine, so there does not appear to be magic in each moment, nor the need to concentrate. Yet meaningful events can happen at any time. To get into flow and find meaningful experiences requires treating each moment as precious. By paying attention to seemingly unimportant events at seemingly unimportant times, minor coincidences sometimes unfold that lead us along a meaningful path. To be in flow is to allow ourselves to be led by these coincidences and discover meaning in them, noticing constellations of seemingly unrelated events that may relate to a common purpose.

In the children's book *The Phantom Tollbooth*, Alec gives Milo a telescope, saying:

> *Carry this with you on your journey, for there is much worth noticing that often escapes the eye. Through it you can see everything from the tender moss in a sidewalk crack to the glow of the farthest star—and, most important of all, you can see things as they really are, not just as they seem to be.*³⁵

This telescope can also help us see synchronicity where it is waiting to be birthed, even in unlikely places.

My father, a building contractor, once ordered a small piece of plastic with a mirrored surface from a local specialty plastics shop for a job he was working on. He only needed one of these pieces, but he ordered two just in case. When he arrived to pick up his order, the item wasn't ready, and he had to wait while the cutter finished another project and then finally completed his order.

Imagine my father standing there in the shop, bored, waiting his turn. If I found myself in that situation, I'd probably get annoyed. I might check my phone, let my mind wander, and become increasingly frustrated at being late for my next obligation. My father waited patiently and stayed attuned to the situation. When the plastic cutter finally finished, he was grateful to my father for his patience, and he sent my father away with not two but four pieces of the plastic mirror material as a token of his appreciation. My father "looked through his telescope" to see that he could use these two extra pieces to build a kaleidoscope with my daughter the next time she came over. The experience of having to wait for his order to be finished led to an opportunity he hadn't expected and might have missed if he'd been in a different frame of mind. Digging in the shallow soil of everyday life, who knows what treasured ideas we might find?

My research as a physicist focuses on trying to understand the foundations of quantum mechanics in order to uncover the physics of synchronicity. But in asking myself how this work is useful in the world, I have concluded that synchronicity leads to flow. Being in flow is how we dance with synchronicity in a practical yet transcendent way. In the following pages I will show how seemingly unrelated aspects of life, such as authenticity in our relationships, living from the heart, and having a sense of faith in meaningful outcomes can all come from living in flow.

2

EXPECT SYNCHRONICITY AND FEEL FLOW

A few years back, my wife, Dana, and I decided to renovate a wall in our guest bedroom. Dana asked, "Are you sure we aren't getting in over our heads here?" We were planning to hire a contractor to do some of the work for us, so I confidently replied, "What could go wrong?" Well, one week into the job, our contractor made a construction error. For reasons of confidentiality, I can't say precisely what happened, but suffice it to say that five months later I was still neck deep in triage and crisis control, writing an email demanding (in not-so-nice language) that the contractor hand over the permit so we could continue the project without him. He replied that he would be forwarding my email to his lawyer.

How did I get to that point? How did a situation get so out of hand that I found myself doing and saying things totally unlike my normal self? That night, I cried in frustration, and in my desperation I gave up trying to fix the situation. Not knowing exactly what I was doing in that moment, I shifted my attention into flow. I let go of the idea that I had the solution, and I looked straight into infinity and said, whether to a divine being or to my own soul, *Everything I have done has made this worse. Please show me a way to the other side.*

To this day, I am not exactly sure what changed. Yet I can tell you that, with no further difficulty on my part, two weeks later we had obtained the permit, we had found another contractor we trusted, and the project was definitively on its way to resolution. What made the difference that night? Rather than trying to fix the problem, manage my wife, and control the contractor, I allowed myself to *feel* the dilemma I was in. Like a participant in a twelve-step program, I was open and honest with myself instead of insisting on my agenda.

Learning How to Feel

The most important message this book holds can be summed up in one word: feel.

My native cultural identity—that of a white, male Westerner—has placed a strong emphasis on stoic control of the emotions. Many of civilization's greatest accomplishments in the fields of technology, democracy, and finance have encouraged us to live in our heads. Yet nobody is immune to their feelings, and although stoicism can be a powerful source of strength at the appropriate times, it is evident from the widespread incidence of depression, opioid addiction, domestic violence, and even acts of terrorism that our culture of stoicism has also created some problems. In order to address the distressing problems we have created, both in our own personal lives and in the larger world, we will need to be willing to feel with our hearts.

To be human is to face tremendous adversity. On September 11, 2001, we in the United States experienced a devastating loss of friends, family, fellow citizens, and, for many, innocence. On Earth Day in 2010 the United States suffered the worst oil spill in its history, which damaged ecosystems in the Gulf of Mexico and in the entire planet's oceans. On March 16, 2011, a tsunami caused major damage to a nuclear power plant in Fukushima, Japan, releasing large amounts of radioactive material into the Pacific Ocean.

It can feel overwhelming to try to cope with these situations, and I believe most of us find a way to shut them out of our minds—and, more importantly, our hearts. But I think an understanding of synchronicity and flow is coming into our awareness at just the right time. I will make the case that synchronicity reflects back to us the truth of how we feel, and flow encourages us to be honest with ourselves and authentic with others about how we feel.

This is a crucial gift for us in our present circumstances. In my home renovation project, this awareness helped me nurture the project back to health, but more importantly, it rescued me from the self-imposed isolation I felt while trying to manage the project myself. By becoming aware of my feelings, I stopped trying to fix the problems and instead shared my experiences openly with Dana. Rather than striving to control the circumstances—which caused pain not only for me but also for the other people involved—I was able to share my struggle with Dana, who had also been suffering through the same circumstances.

Our grief allows us to let go of what is holding us back from living fully authentic lives. Our sorrow, and the outrage that may come with it, can light a fire within us. We don't all need to become activists. We don't all need to follow an outward path of political or social change. Rather, allowing those authentic feelings into our hearts can give us the courage to be honest right here in our own lives, in our own homes, in our own places of work or worship. When we authentically feel our emotions, we are no longer willing to let life go by without speaking up into the microphone. We are also no longer willing to let external standards determine who we are. An important first step is allowing ourselves to feel the grief of loss—whether it's the loss of a relationship with a person we loved, the loss of ecosystems or creatures, or the loss of a profession, a way of life, or an opportunity. When we open to the full range of our experiences, we can also reconnect with our ability to feel joy, to feel gratitude for what we have, and to sense our power to direct our own lives.

Our feelings, according to the theory of synchronicity and flow that I present here, are the interface between us and the world out there. When we start taking responsibility for our feelings, allowing ourselves to experience them—and, when necessary, to heal the wounds they're connected to—I suspect we will find that the cosmos responds constructively. Events will occur that encourage us to rebuild a world that works for everybody. With the clear heart and open mind that come from living in flow, we will "anticipate qualitative experiences" (a phrase we will define shortly) that align with generosity and forgiveness rather than greed or retribution. By doing so, we will increase the likelihood of useful or constructive meaningful coincidences—i.e., synchronicities—occurring.

Meaningful coincidences rely on the notion that some circumstances are meaningful to us, while others are not. To understand this concept, let's think about what we mean by this word "meaning."

Objective Meaning and Subjective Meaning

In order to help us better understand synchronicity, I need to present some new ideas from my own research—ideas that are still in development and are not yet firmly established in the scientific community. To most physicists, meaning is a psychological concept, related to the personal values we place on something.

If you pass me in a dark alley and reach out to shake my hand, and if it's also the case that just last week somebody stole my wallet in a dark alley, I will probably be scared of you. Although your gesture was motivated by kindness and I was perfectly safe, my personal psychology led me to interpret the situation in a certain way. This is what I call "subjective meaning."

Physicists try to avoid issues of personal interpretation like this. That's where "objective meaning" comes into play. This type of meaning can be precisely defined so that it's not a matter of personal interpretation, which will allow us to see how synchronicities might arise.

From Killer to Healer

In World War I, armies made extensive use of mustard gas as a weapon, with horrific effects. Mustard gas attacks the mucous membrane and skin upon contact. As luck would have it, two alert doctors treating soldiers exposed to mustard gas in the trenches noticed that their patients exhibited a decreased white blood cell count. Eventually this astute observation led to the development of mechlorethamine, a drug that inhibits the growth of cancerous white blood cells in patients with lymphoma.

To express this idea, we need to talk a little bit about what it is like simply to be alive. I agree with philosophers such as David Chalmers who use the term *qualia* to refer to the subjective qualitative experiences we undergo as humans.[36] For example, the experience of seeing the color blue, tasting a cherry, or feeling companionship with a friend are all instances of qualia. You will notice that qualia cannot be defined in terms that would make sense to someone who has never had that experience; it would be impossible to define "blue" to a person who has been blind from birth, for example.

Chalmers says qualia give rise to what he calls the "hard problem" of consciousness: what gives us our feeling of what it is like to be alive? We can certainly describe the process of eating a cherry in terms of which nerves are activated in our tongue and brain, but reading that description in a book could never convey to you what the experience is actually like. Similarly, imagine a scientist who has been raised indoors in a purely black-and-white environment but who also knows everything there is to know about the science of color. One day she steps outside into her yard on a sunny morning for the first time, and only then is she actually experiencing color. Philosopher Frank Jackson used this thought experiment to illustrate that there must be something more to qualitative experience than just the physical neurons firing.[37]

It seems evident that qualitative experiences must be something fundamental, something absolutely vital to the experience of feeling alive. It's impossible to imagine what life would be like without qualitative experiences; there would be no *there* there. In Chalmers's terminology, a person without qualia would be a "zombie." But of course, you know what it is like to be you, so you are not a zombie. You are a conscious, living being; thus, qualitative experiences are the foundation of who you are.

So I will take the view that the world is made up of *experiences*. Physical particles like electrons (the tiny particle that conducts electricity) arise to create things—blue cars, sweet cherries—that allow those experiences to happen. Interestingly, physics seems to tell a similar story, if you look at it the way some physicists do. Quantum mechanics tells us there are no objects in themselves, only collections of properties that we associate with objects. The mass, charge, spin, position, and momentum of an electron are properties we can measure, but there is nothing else. What's more, such properties only exist relative to an observer. Without defining a recipient of the information, an "experiencer," there is no definite property for a thing.

Again we can say there's no *there* there. What is an electron? Just a predictable collection of properties. What, then, are experiences? They, too, are a collection of properties. The experience of eating chocolate combines a taste property, a smell property, and a feeling property in the body. Although there's more room for variety with the description of chocolate than with the description of an electron, if you repeated the set of properties associated with eating chocolate in a variety of circumstances, they would all feel quite similar to each other, creating a single basic type of experience. We have experiences based on the properties of the world we measure around us, and those properties are only defined from our perspective. Without the experiencer, no clearly defined properties exist. There is no difference between the properties of things and our experience of those properties, for these two are inseparable. Hence, the world is nothing more than our experiences of it.

In this framework, *the objective meaning of our actions has to do with what types of qualitative experiences our actions lead toward.*

To illustrate objective meaning, consider Liu, who loves having a clean home. Liu spends much of his free time caring for the look and feel of his house. We could say that his actions have meaning, in the sense that they lead to an experience he wants to have, i.e., living in a clean environment. That overall experience is made up of specific experiences that Liu is seeking, such as a feeling of freedom in the kitchen that comes from being able to make a meal without having to clean up yesterday's dishes, or a feeling of calmness that comes from sitting in an organized office.

Usually this intention of his will lead him to do things that actually result in the experience of a clean home. This direct relationship between his actions and their consequences is objective meaning. If he instead spent his free time riding a bicycle, we might infer that he is seeking the experience of feeling physically healthy or spiritually connected with nature, and we could tell this because more experiences would tend to happen where he felt healthy and vibrant. If he spent his time shopping for clothes, we might infer that he is seeking the experience of feeling good about how he looks, and this would lead to experiences in which he feels this way.

It could also be true that actions leading to the experience of having a clean home could arise from a different set of intentions. For instance, maybe Liu cleans his home because he is afraid of being criticized by his mother for having a dirty home. He is choosing what he does in order to have a certain qualitative experience, whether that is the feeling of cooking in a clean kitchen or the feeling of his mother saying something positive about him when she visits. The qualitative experience Liu is choosing may not be immediately apparent, even to Liu himself. Cleaning his home could lead to a number of different experiences, and we define the objective meaning of Liu cleaning his home by the experiences that ensue from it.

Let's say Liu's motivation is really about trying to impress his mother. This may lead to different consequences because he may make subtly different choices, such as being dissatisfied with the effectiveness of product A for cleaning his floors and thus opting to use a stronger product B that he hasn't tried before. If it turns out that product B is so strong that it damages his floors, he may become upset, which may make him suddenly aware of his unresolved issues with his mom. We see here two quite similar actions with differing objective meanings. Being satisfied with product A leads to the outcome of a clean home and reflects Liu's intention to have a clean home. Using new product B because it's stronger than product A causes damaged floors and reflects his need to impress his mother, leading to a situation in which he must confront his mother or at least deal with his hidden feelings. His actions had objective meaning, which we can learn about from their consequences.

I have seen this as a parent. The objective meaning of my actions is generated by their consequences, regardless of my conscious intent. Take the example of my daughter's leg warmers from chapter 1. I may think that by encouraging her to change to matching clothes I am helping her learn to feel better about herself, but my actions are actually undermining her autonomy. The outcome, whether I consciously intend it or not, is that she has an increased sense of self-consciousness and a diminished trust in her ability to choose. This reflects the objective meaning of my action. In this case, identifying the objective meaning of my action required understanding not only the physical circumstances but the emotional and psychological circumstances as well.

Our hidden motivations can muddy the waters and make it seem like all meaning is hopelessly subjective. But just as Galileo realized that a hidden force, friction, caused things to slow down and veiled the natural law of inertia, I believe our hidden motives inform our actions and obscure the responsiveness of the cosmos. I suggest that all actions carry objective meaning, but whether they align with our conscious intent is a different matter. The misalignment between our conscious intention and the underlying reality of the situation can make it hard to accurately see the meaningful connections between events in our lives. Gary Zukav, author of *Seat of the Soul*, says: "If you have conflicting intentions, you will be torn because both dynamics will be set in motion and oppose each other.... The strongest one will win."[38]

The trick to aligning with flow and synchronicity is to be able to identify the objective meaning of our actions and then make adjustments to align them with our conscious intent. Whereas subjective meaning comes from linguistically framed ideas and opinions about experiences—who said what about whom, what do I think about such and such—objective meaning is couched in a "language" made up of experiences themselves. I will describe this concept in more detail soon, but first it will be helpful to explore Carl Jung's work on archetypes and the collective unconscious.

Jungian Archetypes and Symbolism

Jung believed that our subconscious minds include an impersonal level that we all share, called the collective unconscious. He drew this conclusion from a variety of sources, including examples of common symbols used across cultures and at unrelated times and places. For instance, he describes a schizophrenic patient of his who spoke to him about looking directly at the sun and seeing a phallus hanging off of it that created the wind. Jung was of course puzzled by this delusion, but four years later he came across a book about the ancient Mithraic religion in which the same symbolism is described. Being completely sure that there was no way for his young patient to have known about Mithraism, Jung felt compelled to wonder if the young man was drawing on a common subconscious symbolic language. Could there be symbolic forms that have some kind of objective meaning for human beings, no matter who they are? Jung says, "I must emphasize yet again that the concept of the collective unconscious is neither a speculative nor a philosophical but an empirical matter. The question is simply this: are there or are there not unconscious, universal forms of this kind?"[39]

Jung calls these universal forms "archetypes." He explains, "The concept of the archetype ... indicates the existence of definite forms in the psyche which seem to be present always and everywhere." This perspective implies that some symbols convey meaning to us not because of our own past experiences but because of our shared collective experience that "does not develop individually but is inherited."

Another example of a Jungian archetype is the "dual mother," a theme that appears in various forms from Greek myths to the rebirth of Christ through baptism to the twice-born god-kings of ancient Egypt. One could say the dual-mother archetype appears today in the concept of a godmother or godfather. Because there is no biological necessity for humans to have two mothers, Jung says "one cannot avoid the assumption that the universal occurrence of the dual-birth motif ... answers an omnipresent human need which is reflected in these motifs."

My approach to meaning and symbolism is a little different from Jung's. I don't examine the collectively defined symbols themselves, such as the dual mother described above; instead I focus on how our actual experiences symbolically reflect the meaning of our choices. Yet the two views both involve a collective symbolic language in which experiences stand for some particular meaning. In both cases, the meaning of the symbols is determined by the inner state of the person; yet there is a connection between the inner state and the outer world.

What does it mean for an experience to be symbolic? A symbol is something that stands for something else. In addition to thinking of an experience literally in terms of its specific details, we can also think about its symbolic nature by considering the inner cognitive and emotional experiences that result from it. For instance, the experience I described concerning my daughter's leg warmers involved a literal set of circumstances in which we were in our living room, it was early in the morning, she was getting ready for school, and I was questioning her clothing choices. In a way, this experience is similar to an experience we had the day before in which she was getting a snack in the kitchen and I commented on what would be the healthiest choice. The events were pretty different in their literal details, but they were similar in the underlying cognitive and emotional message she experienced: me questioning her judgment. Hence, these two different events "stand for" a similar qualitative experience.

The idea that the physical world is secondary to the world of ideas or meaning has shown up in various forms throughout history, including the analytic psychology of Jung; the philosophy of idealism as expressed in the writings of Plato, Berkeley, and Kant; modern cognitive science in the work of Donald Hoffman and others; and many of the Eastern spiritual traditions. The breadth of possible versions of this idea, and the detailed arguments that exist among their advocates, would take many pages to explore and would lead us away from our main purpose, so I will not pursue them further here.

I come to this view not through my philosophical preferences but out of practical experience. In my understanding, quantum mechanics is very clear that objects do not have definite properties unless we interact with them. Furthermore, in the final two chapters I will make the case that light is "timeless." This means that quantum mechanics applies to everything and that the properties of things we interact with are not the true properties of the things in and of themselves. Rather, we experience properties only relative to ourselves, so there is no objective and definite world "out there." There is only the set of relative, mutually consistent experiences we are each having. I believe these conclusions will eventually be physically verifiable, although this has not been done yet, and there is still great debate about the interpretations of what quantum mechanics really says about the world. Taken together with my own practical experiences of synchronicity, I have come to suspect that objects in the physical world are symbols that represent experiences.

The Synchronicity of Biochemistry

Biochemistry often benefits from synchronicity because the number of possible combinations of atoms in organic molecules is enormous, and the few combinations that are useful are hard to find. The *Journal of Biological Chemistry* has published more than forty articles over the past thirty-five years that include such phrases as:

- "Chance and serendipity deserve co-authorship."
- "[The discovery] involved a large component of both serendipity and insight."
- "Serendipity in science is a wonderful thing."
- "Chance favors the prepared mind—from serendipity to rational drug design."
- "Intermediates in Crystals: Search Long Enough, and They Will Find You"
- "[Thanks to my funder for making] it possible to follow the path of serendipity on which science thrives."
- "Deoxyribozymes: Selection Design and Serendipity in the Development of DNA Catalysts"

A conscious being who eats a piece of chocolate feels an inner experience of *chocolate* and "communicates" with the surrounding physical world via that symbol. In other words, the literal experience of eating chocolate relates to the set of inner experiences of eating chocolate. But those inner experiences don't relate only to that specific instance of eating chocolate; they are related to all the other possible situations that could involve literally eating chocolate under different circumstances. This is similar to how the word "chocolate" describes the specific dark chocolate I am eating now, as well as the milk chocolate I ate yesterday and will eat tomorrow. *The qualitative experiences of life are like a language being spoken between the responsive cosmos and the living creatures in it* (although I will not try to clarify here what distinguishes a living creature from the rest of the cosmos). The inner experience of the organism is the content of what is being communicated, and the literal experiences—the physical circumstances—are the medium of communication.

So the qualitative experiences that happen as a result of our actions lead to a sense of the objective meaning of our actions. From objective meaning we can now understand how what I call the "tree of possibilities" determines the likelihood of future circumstances and can lead to synchronicity.

The Basic Model of Meaningful History Selection

The premise of meaningful history selection is that the world we experience in everyday life is a specific set of circumstances drawn from a multitude of possibilities. The possibilities that could come to pass can be visualized as existing on a tree. (See figure 1.) Our lives get shaped in different ways depending on how we navigate the tree.

Expect Synchronicity and Feel Flow

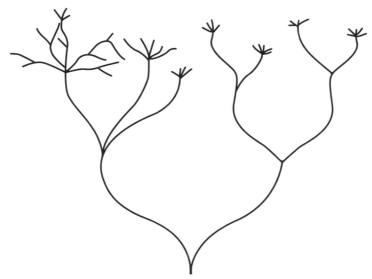

FIGURE 1. The events we experience in everyday life can be represented as a branching tree of possibilities. Each branch represents a different way the world could unfold, and every branch contains the whole universe but in different arrangements. As we move from the bottom of the tree up toward the top, we make choices that determine which branches we end up on. Although every branch represents the whole universe, some branches may have more favorable circumstances than others.

This image comes from the mathematics of the theory of quantum mechanics. The mathematics of the tree are very well established in mainstream physics. However, the meaning of the tree—for instance, whether the branches actually represent parallel universes—is a matter of ongoing debate within physics, which I outline in the later chapters of the book and in appendix B.

We must be very careful when we think about this tree of possibilities. It is not simply a "decision tree" or flow chart to help us maneuver through life as if we were running a business or troubleshooting an appliance. Each branch of the tree represents *all of the universe* branching into different versions of itself. In other words, it is a *decision tree through possible experiences*. Rather than using the tree to navigate our way through a *single* landscape, we use the tree to navigate from one landscape to another. On one of the top branches I may end up as an architect, while on another branch I end up as a sculptor. I am represented on both branches but in two different sets of circumstances based both on my decisions along the way and upon the response that the universe provides at each step. This principle works the same way at a smaller scale: on one branch I head to the grocery store and accidentally run into an old friend, whereas on another branch I choose instead to go to the gym and meet someone new who becomes a collaborator on my next professional project. Either branch can be great, but they involve different circumstances.

A key characteristic of the tree is that if I choose to go to the store and end up meeting my friend, the possible experience of me meeting a collaborator at the gym doesn't exist on that branch. On that branch, the gym may have been closed that day. We can only speak about the branch we are actually on, a concept known as "counterfactual indefiniteness." Other branches, which we might think of as missed opportunities, don't have any "realness" to them. In this model, regretting missed opportunities doesn't make any sense, because opportunities only unfold in response to the choices we actually make, not the ones we *could* have made.

We can think about this tree of possibilities as an apple tree. (See figure 2.) Some branches have apples, and some do not. Some apples may be bigger than others. The apples represent outcomes that align with our choices. In the best case, our conscious and subconscious intentions are aligned so that the outcomes we get are the ones we want. In many cases, though, our choices may be aimed at things we don't really want because we are acting from unconscious motivations that don't serve our best interest. In either case, the apples represent the meaningful outcomes.

The point of meaningful history selection is to navigate through the tree toward apples that reflect our intent. To navigate in this way, we don't just choose where we want to go and go there. After all, the branches on the tree don't represent actual places. They represent *situations*. Moving around the tree involves making decisions and acting upon them based upon our thoughts, feelings, and emotions (in this model there is an

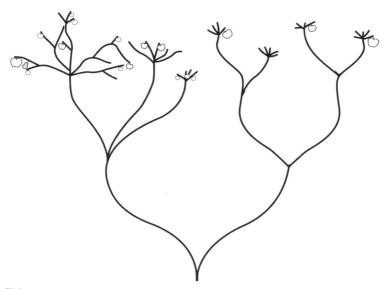

FIGURE 2. Every branch on the tree represents a different way circumstances could unfold. Branches with apples represent circumstances in which a qualitative experience occurs (e.g., a friend invites you for a walk on the coast) that matches your own anticipated qualitative experience (e.g., you yearn to feel some fresh air). The more apples a region has, the more likely it is that experiences in that region will take place. Therefore, your intent (loosely speaking) becomes more likely to come true.

important distinction between feelings and emotions, which will be discussed more explicitly in a moment). Our bodies, I suggest, are designed to anticipate qualitative experiences. For instance, we might anticipate the experience of connecting with a friend. This is a mix of thoughts ("We want to invest in our friendships"), feelings (the excitement of feeling connected to someone we trust), and emotions (the visceral relief from loneliness). We can't help but anticipate future experiences, whether consciously or unconsciously. A multitude of thoughts in each moment trains our attention on our experiences and how we want them to change or stay the same.

But it is not specific experiences that our body tunes into. Rather, it is the *qualitative nature* of those experiences. We may be thinking consciously of a specific experience like meeting a friend at the grocery store, but our bodies are having feelings and emotions that are associated with many possible alternate versions of that experience.

For instance, let's say you are feeling lonely and thinking about how nice it would be to see your friend Anne. When you leave the house, you are deciding between going to the store or going to the gym. You end up heading to the gym, but there is a traffic detour on your way to the gym that reroutes you near the store. You go with the flow, and while at the store you run into another friend, Maggie, which gives you the experience of connecting with a friend.

The theory of meaningful history selection describes this situation by suggesting that your anticipation of the experience of "connecting with a friend" made certain branches of the tree more likely. One branch involved a detour that made it more likely that you would end up at the store. From there, another branch existed where Maggie also happened to be at the store at that time. Even though you were imagining Anne, it may be that there were no reasonably accessible branches in which you run into Anne (perhaps because she was traveling out of the country, for instance). Yet your anticipation of the experience of connecting with a friend helped move you onto a branch of the tree where the same qualitative experience occurred in a different form from what you expected. It is "qualitative" in the sense that it didn't happen as you *literally* expected, but it did match the qualitative nature of your intention. On different branches there were different opportunities for you to have the experience you were after. Your choices shaped the specific way it unfolded for you.

At its root, the theory of synchronicity presented here relies on an association between the properties of things we can measure in the world and the experiences one can have in the world. In other words, the qualitative experiences we anticipate align us with certain types of outcomes—driving recklessly may highlight apples on the tree associated with a cat running in front of our car and resulting in a car crash—and those *types* of outcomes are made of actual possible configurations of real things (or properties) in the world. This association between properties of things and experiences we have, although intriguing, is still speculative and remains a point of uncertainty in the theory.

A New Worldview

I feel that this new worldview can have a profound impact on our collective experience. Recognizing and understanding the way in which our choices shape our world is something we are only barely able to do in our present civilization. We are good at recognizing how our choices shape systems outside of ourselves, such as the economic marketplace or (for most of us) the global climate. But when it comes to human nature, especially our own, it is quite difficult to acknowledge and take responsibility for how the world outside may reflect our choices.

I think it's important not to misinterpret this worldview as a "karmic retribution" approach, where those who are suffering must have deserved it based on past mistakes. The world is more complex than that, and ultimately we are each on our own journey and have to rely on inner knowing to interpret the meaning of events in our lives for ourselves. What this worldview says is that we have influence over the small coincidences that happen in our lives, and if we manage them expertly through the process of flow—with a balance of assertiveness and receptiveness, controlling and allowing—then we can use the responsiveness of the cosmos to build our lives toward our anticipated best possible life. We can also use the responsiveness of the cosmos the other way to sow contention and discord in our life.

We all start out with different circumstances. Those who find themselves in difficult circumstances for whatever reason may have to work harder to get to a stable baseline, but I suspect that it can be done by using the process of flow that emerges from the synchronicities of life. This is certainly not easy, though, because it relies on the hardest thing of all: self-knowledge. In order to make consistent decisions in harmony with our lives and our communities, we must be able to feel our feelings accurately and relate to others authentically. If we don't, the process of meaningful history selection will respond to the underlying feeling—perhaps a desire for self-sabotage or revenge, for instance—and we will be frustrated by the events that emerge in our lives.

If your choices shape your world, then instead of feeling regret about opportunities you missed, it makes more sense to reflect on what you might learn from the situation and how you might choose differently next time. Circumstances support the path you did choose, not the path you didn't choose. In the earlier example of choosing between the store and the gym, since you went to the store, it doesn't make any sense to talk about what would have happened if you had gone to the gym. The world unfolds around the choices you *do* make, not the ones you could have made. If you hadn't chosen to go to the store, you can't say for sure whether Maggie would have gone or not. And since you *did* actually go to the store, it makes no sense to regret that maybe you should have gone to the gym because you need the exercise. On the branch you *did* choose it may be that the gym was actually closed, so it may be a good thing that you didn't waste your time. This set of circumstances unfolded in response to your choice. In this model, you are not "tuning into the fact" that the gym was *actually* closed. Instead of trying to read minds or predict the future, this is about seeking inner clarity and acting in the world with integrity, confidence, and vision so that the cosmos can respond to your highest ideals, not your worries and fears.

If your choices shape your world, then instead of thinking your choices don't really matter in the big picture, you might realize that important things happen wherever you choose to go. Be on the lookout for those little nuggets of joy wherever you are, in whatever you are doing. In flow, new connections tend to happen in unexpected ways, and they always occur in response to the choices we make. Just because you are at a friend's wedding doesn't mean you shouldn't be on the lookout for a new business partner. Just because you are at work doesn't mean you shouldn't be tuned into new ideas for your creative projects at home or ideas for family vacation. Every moment can bring useful information, and how you respond to the nuggets of joy that show up randomly throughout your day makes a huge difference in what happens next.

If your choices shape your world, then instead of thinking the barriers in front of you are insurmountable, you will look for the coincidental clues coming from the cosmos, the chinks in the armor that might allow you to gracefully turn a problem around. Even if the problem you are facing exists on most of the branches on the tree, if there are some branches of the tree where the problem does not exist, then there might be a way to get there. By seeking alignment between our conscious desires and our subconscious motivations and looking out for surprising opportunities, we may move ourselves closer to the outcome we want.

If your choices shape your world, then it is pretty difficult to feel insecure about your importance in the world. After all, if the cosmos is responding to you, then you are important! You may not feel happy with your circumstances, and you may feel insecure about your talent, preparation, intelligence, or appearance, but you can't possibly feel unimportant. The cosmos is forming itself around each of us, and that's a crucial role we play.

If your choices shape your world, then it is easier to find harmony in your personal relationships. Other people and their choices serve as a part of the circumstances supporting your path. If their choices conflict with what you want, that is a chance for you to reflect upon why you think your success or happiness depends on something outside yourself. For instance, let's say you have a business partner who is leaving the partnership. Will that stop you from proceeding toward your own goals? In this new situation, might you be able to accomplish some kind of inner growth—such as developing the confidence to go it alone—that will allow you to keep taking intentional action? If you do, the responsive cosmos will continue to send you small opportunities to build your business, whether your business partner is there or not. Living in flow and paying attention to synchronicity can help us develop healthy relationships. Since the cosmos is always responding to us and is a reflection of our choices, we are fully empowered in and of ourselves. Other people serve as joyous opportunities to collaborate and to learn about relationship, and they cease to be sources of pain, frustration, and irresolvable conflict.

If your choices shape your world, then you might begin to feel that the only basis for evaluating your decisions is to question whether they are really what you want or not. Each choice is the "right" one within some frame of reference. Each choice is building momentum toward some anticipated experience. You must know yourself well enough to know whether the experience you are building toward is the one you want to have. Instead of questioning your decisions and looking outside yourself to determine whether you have made a good decision, you will see exactly what type of experience your action is building toward and know right away whether it is what you intend.

Feelings Drive Synchronicity

I am now going out on a limb, beyond my area of scientific expertise, to tell you what I think this all means from my experience.

We are beings of experience living in a world of emotion. We are immersed in a sea of subconscious affect or feeling, and it is our feelings that drive the responsive cosmos. Our feelings pull us toward (or pull toward *us*) the meaningful events of our lives.

What do I mean by "feelings"? And how do they fit into meaningful history selection? The model I will use here to define feelings is drawn from research in neuroscience, psychology, and philosophy. It distinguishes between experiences, emotions, feelings, and thoughts.

Psychologist Paul Ekman did groundbreaking work in the last half of the twentieth century on categorizing human emotions.[40] Ekman is perhaps most well-known for categorizing the facial expressions associated with human emotions and identifying their universality. Through his work it became clear that emotions are not only a learned cultural response to our experiences but are, to a large extent, universally programmed into us.

The work of neuroscientist Antonio Damasio[41] further illuminates the distinctions between emotions, feelings, and thoughts. According to Damasio's somatic marker hypothesis, emotions play an important role in efficient human reasoning and decision making. Philosophers like David Chalmers[42] support the notion of qualitative experience (or qualia) as a fundamental concept, separate from the thoughts, feelings, and emotions we have.

In the phenomenological model I'm using, everything starts with having experiences in the world. Emotions are then an automatic physiological response that our bodies have to our experiences—what Damasio calls the proto-self. Stomach butterflies when we are afraid or goose pimples when we are in awe are bodily reactions to emotions that we don't have any conscious control over. Feelings, on the other hand, are how we *perceive* the emotions. Having feelings gives rise to what Damasio calls core consciousness. Feelings are more complex than emotions because they depend on how we perceive our circumstances. A seasoned performer may respond with feelings of joy and excitement to the emotion of butterflies in the stomach, whereas someone with stage fright may try to run away as far as possible. By adding our own mental interpretation to our emotions, the five basic emotions of fear, anger, joy, sadness, and disgust become a broad field of possible feelings or interpretations. Finally, our thoughts are those connections we make "in our head" while trying to make sense out of our experiences and our emotions.

Taken together, these considerations suggest that our life consists of experiences, to which we have emotional reactions, which we interpret as feelings and which lead to cognitive thoughts about life. (See figure 3.) This isn't a comprehensive picture of the human condition; it's just a simplification of what we know from cognitive science, psychology, philosophy, and neuroscience to help me (a physicist) build a model for how meaningful coincidences show up in life.

"Feelings" are more than just the acute expression of emotions, such as feeling angry when my spouse forgets to meet me at the bus or feeling frustrated when I leave a bag of groceries at the store. Rather, feelings are ubiquitous in everyday life whether we notice them or not. They decide where we want to go out to dinner, what movie we want to see, or which business presentation we liked better. Without our feelings, we might be unable to make decisions. We eat at a Chinese restaurant simply because we "feel like it." Our feelings tell us.

FIGURE 3. We are creatures having experiences, to which we have emotional reactions, which we interpret as a variety of feelings, triggering thoughts that seek to make sense of the world. In reverse, our conscious thoughts can influence our feelings, which can regulate our emotions to some degree, which can (according to meaningful history selection) influence the types of experiences we have.

> ### Following the Flow in Graduate School
>
> Contrary to my normal approach, I decided to do graduate school "by the book" by taking the classes in the recommended order. This meant signing up for a class I was not excited about, but I would get it out of the way. As it turned out, by some fluke the class had only four people in it, and it ended up being a low-stress environment for learning the material. I ended up loving the course! The next year it was offered, there were twenty people in the class, and my friends who took the course told me it was a real bummer.

But feelings are more than just a data point. They *drive* us. Even with all the external data pointing in one direction, our feelings can make us choose another direction. Have you ever ignored what seemed smart or reasonable because you felt drawn to something else? Have you been drawn to somebody who was not at all your type? Feelings have a power behind them that rationality lacks. We go in a certain direction because we are drawn by the experiences we feel in our body.

Zukav writes, "Every action, thought, and feeling is motivated by an intention, and that intention is a cause that exists as one with an effect. If we participate in the cause, it is not possible for us not to participate in the effect. In this most profound way, we are held responsible for our every action, thought, and feeling, which is to say, for our every intention."[43] If we associate Zukav's word "intention" with the phrase "anticipated qualitative experience," his words describe how our actions contain within them the thoughts and feelings that move us toward specific areas of the tree of possibilities.

So what then are "experiences"? What philosophers call qualia are the supposed raw felt experiences of being alive. As living creatures, we are always either having experiences or seeking experiences. Think of philosopher David Chalmers' question, "What is it like to have a certain experience?" Put in more concrete terms, what is it like to drive a car? Picture yourself behind the wheel of an automobile. There is a certain quality of life experience associated with moving fast over a curvy road, feeling the comfort or discomfort of the seat, tuning the radio to your chosen station. These experiences cause emotional reactions and feeling responses that may cause you to yearn for the experience or dread it.

However, you can never directly convey to me the *experience* of driving a car. You can only remind me of similar things I myself have done. You could tell me that driving a Ferrari is like riding on a zip line or flying over the ocean in a glider, but you can't convey to me the actual experience you have of driving the car.

Qualia cannot be captured and shared; they can only be experienced uniquely by each of us. We are not robots creating printouts of data collected from our senses. We actually have a feeling of the experience of being alive. Without experience, emotions and feelings would fall flat. When we have a feeling, it is compelling to us because it ultimately relates to an experience we have had or will have.

This is where meaningful history selection comes in. I suggest that we influence events by associating certain feelings with certain experiences. Not only do experiences lead to emotions, then feelings, then thoughts; the chain goes the other way, too. (See figure 3.) When we have feelings that are associated with certain experiences, we are "tuning into" those experiences on the tree of possibilities. The qualia associated with the experience of driving a fast car can be represented as one or more branches on the tree of possibilities. Our feelings (which I will use as shorthand for all three: thoughts, feelings, and emotions) are constantly anticipating qualitative experiences that we are drawn to or feel aversion to. The inherent anticipation we might feel for the experience of driving a fast car puts apples on any branches of the tree that include those qualia; then, meaningful history selection influences the probabilities on these branches.

Feelings pull experiences to us like water is pulled to the sea. It may not be clear which path water will take, but it adapts to the terrain and eventually arrives at its destination. In the same way, meaningful experiences are drawn like rainwater, one way or another, down the mountainside toward us. We may not be able to predict which actual circumstance will occur, but we can count on it reflecting the nature of our inner experience.

Our feelings drive the events in our lives.

The exact form of each experience we draw to ourselves is unpredictable, but the circumstances are likely to reflect the feelings behind our action. If we are struggling with the temptation to eat dessert or are avoiding a physical attraction toward someone, an inner part of us may still be seeking a feeling we will get from the "forbidden" experience. The qualia associated with eating cake are like hooks in our consciousness. The imagined experience of rich chocolate frosting and the crumbly texture of the cake (Freud's id comes to mind) transcend any particular civilized thoughts of constraint we might have about them (Freud's superego). They are the raw ingredients of life that we yearn to taste—not data outside ourselves, but inner experiences we seek to have, consciously or not.

Then, through meaningful history selection, that feeling of attraction toward a symbolic experience aligns with certain qualia on the tree of possibilities. Any number of potential situations could arise that align with this feeling—we end up being offered a piece of cake, or we bump into the person we're attracted to unexpectedly—and these particular experiences become more likely to happen. Our feelings, conscious or unconscious, guide the flow and shape our worlds accordingly.

This seems to align with Damasio's view of what he calls intention or the "life urge." He says, "Being a brainless and mindless creature, an amoeba does not know of its own organism's intentions in the sense that we know of our equivalent intentions. But the form of an intention is there, nonetheless, expressed by the manner in which the little creature manages to keep the chemical profile of its internal milieu in balance while around it, in the environment external to it, all hell may be breaking loose."[44] He emphasizes that there is something universal about this life urge, something separate from what we humans call consciousness. He says the urge to stay alive "is not a property of humans alone. In some fashion or other, from simple to complex, most living organisms exhibit it." For our purposes here, I am associating the combined interrelationships in figure 3 with a life urge that I call the "anticipated qualitative experience."

So where do the apples on the tree come from? They hang from the branches whose circumstances match the qualitative nature of the feelings we are having (or seeking to have). In the model, the size of the apples appears to be influenced by the emotional intensity of our action: the stronger the emotion, the larger the apples on a given branch.

Hidden Feelings Matter

If feelings drive the events in our lives, and emotions amplify the intensity of that drive, what impact do unconscious or hidden feelings have? We all have unconscious emotions and feelings, and their power remains untamed below the surface. Reporting on their studies of communication in the workplace, Patterson and colleagues say that "the worst at dialogue fall hostage to their emotions, and they don't even know it."[45] The force of hidden feelings can be frightening to our conscious mind because we know instinctively that something we can't control lies down there. For instance, when somebody close to us dies, an uncontrollable wave of grief can arise inside of us that completely mystifies our rational, conscious mind. We come to associate our emotive power with primitive, uncivilized, and uncontrollable urges.

Songwriter and inspirational speaker Karen Drucker tells a story of standing in a supermarket aisle as a grown adult and suddenly breaking down in unrestrainable tears of grief over the death of her mother. A store employee comes along to ask what is wrong, and Drucker says she'll be okay, she is simply crying because her mother died. The employee says, "Oh, I am so sorry, when did this happen?" Drucker replies matter-of-factly, "Twelve years ago."

I find that I can learn a lot about the smaller moments of loss in my everyday life from more intense experiences like grieving the death of a loved one. Feelings of disappointment or fear that I have when things don't go my way are like a mini-death. Even a small disappointment can represent the loss of some small hope that I had, and without me even noticing it, my emotional state shifts. According to this model, I am then more likely to come across experiences that reflect my feelings of frustration or hurt.

In the first year after Dana, Ellie, and I moved to a new city, Ellie's school had a communitywide movie night. Dana and I were looking forward to the opportunity to make some new friends because we had been struggling with feeling isolated since the move. Somehow, though, in a conversation with some other people at school, a difficult topic came up, and we suddenly felt like we had put our foot in our mouths. Had we sabotaged our chances of being friends with these people? Our underlying

> ### Maybe If You Could Just Relax ...
>
> John Cade was an Australian psychiatrist who sought to determine the cause of excitement and euphoria in manic patients. To test his hypothesis that these mood swings were a result of high levels of uric acid in the blood, he had to find a soluble form of uric acid to test on guinea pigs. The most soluble form was lithium urate, which had the unexpected effect of calming the pigs down. Cade synchronistically stumbled upon the therapeutic effects of lithium, which is now used as a standard treatment for mania and related diagnoses.[46]

hidden feelings were fear and self-consciousness about whether people in the community would like us. This underlying narrative was reflected in the event that showed up for us: an "opportunity" to feel really self-conscious. Note that the sour conversation we stumbled into, which I am calling a synchronicity, did not directly solve our problem. Rather, it forced us to face our feelings head-on.

Afterward, Dana and I had a conversation about the experience in which we were compassionate toward each other and more honest about our own feelings than we had been previously. The conversation brought a little healing for both of us. Two days later we found ourselves at a birthday party for a kid in Ellie's class. This time, we each had a stream of experiences where we felt more connected to parents or kids in the school. Dana even followed up with the parent whom we had the sour conversation with, and she found that they were able to move beyond it and get to know each other a little better in the process. This time, the world reflected a different emotional state to us, one that had been eased as a result of the healing conversation we'd had. We were grateful for the change.

The world is always responding to our *entire* emotional state, not just what we consciously intend to feel. If we have hidden emotions, they shape our world because synchronicities appear in alignment with what we feel, hidden or not. The best way to recognize hidden emotions is to look at the circumstances life brings and then check in honestly with our inner knowing. Is there something useful that this experience is bringing us? Is there a way in which this situation spontaneously mirrors our feelings about an issue? Does this feeling seem to keep showing up in our life? This may allow us to become clearer about what the meaning of a circumstance is for us and what our role may be in it.

Medieval scholar and saint Albertus Magnus also concluded that feelings appeared to be a source of synchronicity.

> *A certain power to alter things indwells in the human soul and subordinates the other things to her, particularly when she is swept into a great excess of love or hate or the like. When therefore the soul of a person falls into a great excess of any passion ... it binds things and alters them in the way it wants. For a long time I did not believe it ... [but] I found that the emotionality of the human soul is the chief cause of all these things.*[47]

Jung comments on this passage: "This text clearly shows that synchronistic happenings are regarded as being dependent on affect (e.g. emotion or feeling)."[48] In my experience, feelings become our enemy when we don't know what we feel. *A hidden feeling can cause an unfortunate synchronicity.* By untangling the mess of our feelings—our response to our natural emotions—we can more easily trace the underlying currents of synchronicity, and our emotions are more likely to empower us rather than undermine us.

By contrast, when we are directed by hidden feelings, we might feel like victims of circumstance. We just can't see how our choices may be shaping the situation, and the same frustrating things keep happening to us. Why? Because in the responsive cosmos we are active participators in the unfolding of events. Zukav emphasizes, "It is, therefore, wise for us to become aware of the many intentions that inform our experience, to sort out which intentions produce which effects, and to choose our intentions according to the effects that we desire to produce."[49] Things don't happen *to* us, they happen *through* us. We connect to the future branches of the tree that align with our actions.

This view is not foreign to mainstream physics, although the field does not embrace it in the same way I have explained it. Physicist John Wheeler said, "Every item of the physical world has at bottom ... an immaterial source and explanation ... in short, that all things physical are information-theoretic in origin and that this is a participatory universe."[50]

Physicists are still uncovering the mystery of information as a *thing*, a proposition sometimes restated as "it from bit." It is not that far to go from conceptualizing information as a lifeless description of our "mental state of knowledge" to conceptualizing it as a vibrant reflection of our quality of being.

So although the trivial details of life seem random, I suspect that the overall flow of circumstances in our experience is coordinated. Life is orchestrated by our feelings. However, it is our actions that make all the difference, because our actions broadcast our feelings to the external world. By becoming more self-aware, we regain control of our actions from the grip of our feelings. When we proactively unearth our hidden feelings, we can be freed to choose those actions that shape the world we want, and meaningful events get drawn to us like water flowing down the hillside toward the sea.

GETTING INTO FLOW

It's late at night, and I'm in my bathing suit, headed to the hot tub. I am with a team of people who have come together at a rural retreat center to design a mobile phone app over the course of a weekend. A friend intercepts me in the middle of a courtyard to tell me the hot tub has an "out of order" sign on it, and I moan with disappointment. A complete stranger is standing nearby in the dark, presumably a guest from one of the other groups sharing the retreat center with us. He overhears our conversation about the hot tub and suggests I should come join his meeting. They are having social time and would enjoy the company. It's an unexpected and unusual offer, so I am intrigued. I debate the situation internally a bit before deciding to do it. I change out of my bathing suit and head to the social hall.

I chat with a number of folks sitting in a loose circle, sharing information about my research into flow and synchronicity, and enjoying myself. At the end of the night, only a few of us are still awake, and I find myself talking to a man named Michael. From what others have told me, he is sort of the patriarch of this group. As we talk, he becomes interested in my research and tells me he has a background in physics. He mentions the possibility that I could come speak for a professional association he is part of. We finally head to our cabins for bed, and I find myself very glad for the redirection away from the hot tub and into a very interesting evening.

A month later my good friend Janet from the National Speakers Association calls me from out of the blue. It turns out she and Michael are both on the board of a professional consultants association, and they need a substitute speaker for the following month's meeting. Earlier that week Michael had mentioned my name to her, to which she had replied, "You know Sky too?" It was a done deal.

A month after that I speak to their group, and it proves to be an important event on my own professional path. I make some meaningful new business contacts, I meet a professional designer who offers to help me enhance my slideshow, and I capture great footage of my talk. How was I to know that a broken hot tub would lead to these valuable developments in my career? In fact, I learned later that the hot tub had actually been fixed, but the "out of order" sign had been left in place by accident!

The LORRAX Process

My experience with the hot tub reflects a pattern I think we can cultivate. Jaworski encourages us to listen to the nature of things that are unfolding and then to "create dreams, visions, and stories that we sense at our center want to happen."[51]

To see this process more clearly, let's break my experience down into the following steps: Listen, Open, Reflect, Release, Act, Repeat (X), or LORRAX. (See figure 4.) This cycle of steps can be useful in navigating quick, spontaneous decisions like the one where I decided to adapt to the broken hot tub and join the other group's meeting. It can apply equally well to long-term decisions extending over many months. As defined in chapter 1, synchronicities are meaningful events that seem to defy chance.

Getting into Flow

The LORRAX process optimizes our likelihood of noticing synchronicity when it pops up unexpectedly. When we follow it, we are more likely to get into flow.

The process starts with *listening*. In the story of the broken hot tub, I was standing in the cold (it was November!) and was quite disappointed to find out that the hot tub wasn't working. When a random stranger spoke up about switching gears to come socialize with his group, my normal state of mind would be to ignore this information. Who is this person? What do they have to do with me? I might naturally disregard his comment as irrelevant. The first step to getting into flow is to listen for exactly this type of unexpected information. Useful guidance may come from all sorts of places, not just from places of authority (like the retreat organizers telling us the official social schedule) but also

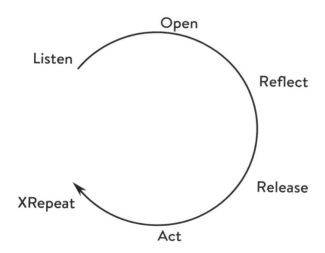

FIGURE 4. The LORRAX process helps us get into flow by fostering both the receptivity to listen to our circumstances and the assertiveness to take action.

from the humblest of sources (like participants in other groups inviting me to join them). To be able to catch synchronicity and get into flow, we have to be able to notice the unexpected opportunities, the calls to a bigger life.

But useful information, by its nature, often conflicts with the status quo. If we only listened to information that matched our expectations, we wouldn't learn anything new. Hence, in order to catch an unexpected opportunity, we may need to *open* our minds to what the information tells us. Listening isn't enough if we have automatic reactions that keep us from digesting new facts. When I received the invitation to socialize, I was standing in the cold in my bathing suit and a towel. The idea of changing back into my clothes and being extroverted was totally different from the relaxing experience I had expected to have. Instead, I wanted to go back to my room and quietly read a book. Yet I decided to remain open and think to myself, *Maybe*. So as I stood there in the courtyard in the cold, I had already performed two steps in the process: I'd listened to the invitation and opened to the possibility.

The next steps are to *reflect* on the situation and *release* expectations. While I changed my clothes, I was still undecided. I thought about how urgent the task of networking had become in my work. I knew from past experience that every opportunity to talk to people about my work was useful because it gave me practice at explaining what I was doing and sometimes even led to further opportunities. I also reflected on the absurdity of the situation. Why was the hot tub broken that particular night? Why had the man directed his statement right at me and not at the other person I was with? It seemed almost like a setup. How could I not go see what might be there for me? By *reflecting* on these factors I became more clear that it was a good idea to follow the opportunity.

Yet I also had an attachment to what I wanted to do. I was still getting over the disappointment about the hot tub, and I just wanted to relax. In order to reflect with an open mind on the situation I had to *release* my attachment to what I wanted to do. While I did want to relax quietly in my room, even more than that I wanted to make new connections that would move my work forward. To follow that possibility, I had to let go of my momentary wish for comfort and surrender to the flow.

Finally I was ready to *act*. I put on my jacket and went to the social hall. Being a person of action is highly valued in today's business environment, yet without the first three steps to inform our actions, we are likely to miss the potential for synchronicity. The first three steps allow us to see more clearly what the opportunity at hand is and what might be done to enhance it. This makes our choice of action more effective. If we don't align with circumstances, we are just imposing our will onto the world and very likely causing problems in the process. The purpose of the LORRAX process is to bring a balance of both receptivity and assertiveness to our decision-making process, or what I think of as a balance of feminine and masculine or yin and yang approaches. It is a dance of aligning with circumstances at the same time as we bring circumstances into alignment with us. We can't just listen, and we can't just act. Both are required.

The *X* in LORRAX is a reminder that life is an endless cycle of these steps. We are always at some place in the cycle. At any time we might be faced with surprising or disappointing information, like a random invitation in a dark courtyard. Ann McMaster calls these specific moments that shake us up "lifeshocks."[52] The lifeshock may be an offhanded comment by a colleague that ruffles our feathers, or learning that our company is going through a major organizational change. It could be our spouse saying they need us to watch the kids on a night when we have a regular get-together with friends, or it could be our child telling us they were teased at school today. Lifeshocks such as these can pull us off our emotional center without us even knowing it. Half an hour later we may find ourselves wondering why our mood is suddenly in a funk.

By listening, opening, reflecting, releasing, acting, and repeating over and over again, I was eventually led to meeting Michael and then to a speaking engagement a few months down the road. Lifeshocks like this can often come in the form of obstacles, like the broken hot tub. Seeing obstacles from the right perspective and following the LORRAX process can help us get into the experience of flow. The transition from listening to opening to reflecting to acting may take place in a minute, an hour, a day, a week, a month, or longer. We can go through the process many times a day for small issues, and we might also have instances where the cycle unfolds over many years.

The cosmos is responsive, and we can be responsive too. In the old paradigm, too often our leaders in business or politics are dead set on their perspectives, so their actions don't get into sync with the situation as it unfolds. When we barrel through obstacles or rigidly hold onto a perspective, we may do unintended damage along the way. Flow allows us to navigate obstacles in a manner that has the most synergy with everyone involved.

It's Not Just 'Going with the Flow'

I believe we are creatures of flow. Flow is natural to each of us in our own way. At a party we may feel engrossed in a conversation and want the rest of the party to disappear, or we may naturally feel the ending of the conversation when one person goes to refill their drink. At work we may feel totally engaged with a task for a period of time and then naturally sense when it has fulfilled its course for the day and we move on to something else.

However, the flow of synchronicity doesn't encourage us to go with the whims of our fancy. Being in flow is not always "going with the flow." Getting into alignment with synchronicity is not a passive practice. Shaping our world often involves lighting a spark. A beneficial set of circumstances—an area of the tree with many apples—may be obtainable, but it may depend upon taking a risk of some kind or pushing back on the status quo. Flow teaches us how to listen to and align with the circumstances, but sometimes we have to go against one set of circumstances to align with another. Through developing and trusting our inner senses or intuition, we can become better at discerning which is which.

For Stephen, the little boy with tuberculosis, to avoid returning to Nazi Germany just before the outbreak of World War II he had to push back against his mother, who did not see the situation as a grave threat. Hitler's eventual invasion of Prague made it clear that Stephen was in alignment with the much bigger flow of circumstances unfolding in the world. In my home renovation project, my original approach to the project had been to go with the flow and trust the contractor to handle everything. When he made a mistake, my response was to take as much control as possible and try to force everything. Both approaches led deeper into crisis. I finally learned how to *get into* the flow without just *going with* the flow. When I saw an opportunity to speak up and help the situation resolve, it was like lighting a small spark. If my perception was on target, then the kindling would catch, and I could step back and allow the situation with the contractor to flow toward any apple that represented resolution. In the hot tub situation, the hot tub being broken was an obstacle that completely disturbed the flow, yet it allowed me to get into a different flow and meet an important new business contact.

Sometimes flow involves letting go of attachment to a certain path; sometimes it involves pushing through resistance to stay on a certain path. Flow is a middle path between rigidity and spontaneity. In flow, we keep our highest goals in mind, and then our instinct helps us interpret the circumstances in the most productive way. How do we know when to push and when to surrender? Flow teaches us to simultaneously fight for what we love and also "trust the river" of life events. I find it helpful to maintain a clear bottom line: what is important to me? What are the "apples" I am headed for? In the case of the broken hot tub, I knew my bottom line was a desire for opportunities to network and practice public speaking. Because I have made consistent efforts overall to find these sorts of apples, I was on the lookout for precisely this type of unexpected opportunity. When my hot tub plans were foiled and I was tempted to go back to my room in disappointment, I remembered my bottom line and figured it was a good idea to follow the guidance presented to me, to see what apples might be waiting. Knowing our core values and goals can help us respond spontaneously to unexpected situations and bring about meaningful results.

Just Give It a Shot

"After work one evening I went to my car, only to realize I had left my keys in my office much earlier in the day. I hadn't used them since my office mate was there when I left. After a moment's despair, expecting that the office had already been locked up, I realized the best thing to do was to go back to the office and see what I could do. When I arrived at the closed door of my office, I confidently put my hand on the knob ... and it turned! My office mate had accidentally left the door unlocked hours earlier, just when I needed him to." (Story contributed by Laura Verrekia)

If we are staying at work late because we are stressed and trying to impress somebody, and in the meantime we are sacrificing our personal relationships and building up resentment, then we may not be in flow. But if we are staying late at work because we are excited about our project and want to work on it, that may be the flow for us at that moment. Sometimes being in the flow requires sacrifice, hard work, and loss of sleep; sometimes it requires ease, relaxation, and surrender. Sometimes building what we love requires pushing a boulder up a hill, and sometimes it involves letting it roll down the hill and adapting to where it stops. Just as with an intense tennis match, extraordinary effort can be part of the flow. How do we know? We can tell because of how we feel when we are doing it. If we are clear about our purpose in what we are doing, we naturally enter into flow in any activity.

This may seem challenging to incorporate into a highly structured situation where the errors have big consequences, such as using flow to meet one's quarterly financial goals and keep one's job. In Frederik Laloux's book *Reinventing Organizations*,[53] he shows us how a number of self-managed organizations incorporate this sort of attitude into their workflow. Organizations that are built on hierarchical decision making often don't have much room for flow. When your manager walks up to you and says, "I need this done right now," there's no room for listening, opening, or reflecting; you just have to act in the way you're told. But the organizations Laloux studied operate in a different way: they are self-managed. The organization feels more like a single organism than a bunch of individuals. Various processes are put in place to ensure that the power for decision-making is distributed across the organization and that each individual has incentive to take responsibility for high-quality outcomes.

For instance, any member of such an organization can make any decision, but they must use what is called the "advice process."

In this process the decision maker must get input or advice from all stakeholders who will be affected by the decision and then consider that advice seriously. They are then empowered to make whatever decision they deem appropriate based on the information they have gathered. The pressure to make good decisions comes from the sense of accountability each member has to their team. When each member is trusted and respected, a sense of coherence within the team grows, which supports accountability.

This approach nurtures flow because each individual is empowered to follow what they deem to be the best course of action in the moment; they are encouraged to trust themselves. Laloux found that for this mind-set to work, it is important for this approach to be fully implemented rather than just paying surface-level lip service to the notion of mutual trust. When people truly know they are trusted, they will feel a sense of ownership and will bring everything they have to their efforts. A synergy and dynamism can emerge that results from a greater sense of flow.

One of the organizations Laloux studied was a power generation company.[54] Obviously a utility company needs strict standards of control in order to ensure safety and reliability, but they are able to do so within a model of self-management. One of the team members was originally from Pakistan and had the idea that the company should build a facility in his native country. The CEO felt it was a risky venture and was skeptical, but the advice process meant the decision was the employee's. After following through on the advice process, the team member decided to go for it. He had at his disposal all the resources necessary to implement such a vision, and he was able to enroll enough of his fellow teammates in the project to get it off the ground.

The ability (or inability) to enroll others in an idea is a natural check and balance that exists within flow. The idea of lighting a spark is to see hidden potentials and ignite them. If there is no kindling around—if none of a person's teammates can see the same value in the idea—that might indicate that the action is not in the flow. In today's dominant system of hierarchies, in which one person's voice has the power to silence the rest, this natural system of checks and balances doesn't have a chance to emerge. Without flow, we rely on obedience to get things done. When we expect obedience of people, we don't win their enthusiasm, and their most valuable asset is withheld from the project.

Allen Combs and Mark Holland, who see synchronicity through the lens of the mythological trickster, warn us that being alert to synchronistic guidance from life "does not mean that we should throw ourselves squarely in the path of coincidence, accepting every chance event that befalls us as some divine gesture. To do so would be foolish indeed, for our benefactor is, after all, a prankster, and enjoys nothing better than making us play the fool. Being observant and alert is more important than blind submission."[55]

Flow is not a structureless, laissez-faire attitude in which anything goes. It is a dynamic balance between will and surrender, a democratization of information that can be facilitated via specific organizational structures.

Healthy Relationships Allow Flow

Relationships of any kind can be hard. We have a deep human need to feel safe, and relationships can be where our safety is most unsure, whether at work or at home. My instincts for self-preservation don't always lead me in the direction of better relationships. I am often attracted to people I *feel* safe with, and I choose actions that *feel* safe with those people. But feelings can be limiting, misleading, and sometimes self-sabotaging. Choosing the safe actions or the safe relationships can lead us to miss important opportunities for growth or success. Instead of shutting down or giving up when we feel unsafe or unseen, that might be just the time to take a risk, be confident in our vulnerability, and show up authentically.

Flow can be useful in our intimate relationships. It allows for nonattachment to specific outcomes because we trust that there are many ways to get what we need on the tree of possibilities. This nonattachment can serve to enhance our intimate relationships by nurturing healthy interdependence. When we are guided by flow, we allow our loved ones to be themselves. Jaworski suggests that "once we see relationship as the organizing principle of the universe, we begin to accept one another as legitimate human beings."[56]

Recently Dana and I have been trying to involve eight-year-old Ellie in more household chores. The hardest part is that we seem to need to continually remind her in order for the jobs to get done. Tensions build, and arguments are common. In this situation, flow has helped us grow to become more effective parents. We become stressed by the situation because when Ellie takes a long time to complete her chores, we are not able to complete ours. Therefore, dinner may get delayed, which delays Ellie's bedtime, and then we don't have time to focus on our own lives before we have to go to bed. Does that sound familiar to any other parents? To avoid this situation, we become attached to things going exactly as we want them to, and we end up needing Ellie to do her job in a very specific way. She feels like we are being bossy, and she resists.

By applying the LORRAX process, we have found a new sense of balance as a family. In listening, we noticed that we were all stuck in a tug of war that wasn't improving, and Dana and I were exhausted from the sheer amount of work needed to keep the family running. We opened to the possibility that our current approach wasn't working and that we didn't completely understand the situation. Maybe we *were* being too bossy! Upon reflection, we realized that Ellie felt as if she had no power to choose her schedule, and we felt as if we had no power to get her to respond. We released our attachment to having dinner together and the timing of bedtime, and then we acted: we started asking Ellie to do the dishes—a bigger chore than any of the others she had done before—and we gave her total autonomy. With the kitchen door closed and her favorite music playing in the background, she could take as long as she wanted to finish the dishes.

What did we get out of it? Well, not doing the dishes, for one! But in addition, the stress we had been feeling from trying to parent via control mechanisms subsided. Dana made a rule that she wouldn't talk to Ellie after 9:15 p.m., so she could be assured of her downtime even if Ellie wasn't observing the proper timeline for bed. We ensured that we got what we needed at the same time as we gave her the autonomy she was craving. We developed more flow in our family, not by letting go of what we wanted but by letting go of the worry about getting what we wanted. In the end, we usually eat dinner together, Ellie usually wears clean pajamas, and she has a reasonably consistent bedtime. Just because we let go of control doesn't mean we didn't get what we wanted. It just comes about in a more organic way, where everyone feels fully empowered and respected.

Being in flow with my daughter allows me to disentangle my needs from hers and exist in healthy interdependence. Just the other night, Ellie called out from the dark thirty minutes after lights out and said, "I forgot to do my exercises. Can I do them now?" My reactionary response would have been *No! Just go to sleep!* But I *opened* to the fact that there might be some benefits to her request. Upon *reflection* I realized that doing her exercises is a great habit and might actually lead her to falling asleep more quickly. I *released* my expectation of how bedtime should go, and I told her she could do her exercises and then go straight to bed, and I didn't want to hear from her again that night. When I checked on her half an hour later, she'd pulled her blanket onto the floor and had fallen asleep right in place after doing her pushups! It was an adorable moment that wouldn't have happened without us getting into the flow together.

The nonattachment of flow can help us navigate the unexpected. If there was ever a place where the unexpected can arise, it is in intimate relationships. My marriage goes better when I can emerge from swimming in the soup of my emotions and ride the waves gracefully. Have you ever disagreed with a spouse or partner on how to carve out time for what each of you needs? In my relationship with Dana, there can be times when the needs of our daughter, our friends, and our family make it hard for us to communicate about what we need for ourselves.

Picture the scene: it is the first week of our daughter's summer break. We returned last week from a brief vacation together, and now Dana and I are trying to keep our professional lives moving. Having a young child home all day means inevitable interruptions. What's more, we have a few new baby cousins in the family whose families need our support, and my brother wanted our help as he tried to find a dog sitter before he went on vacation. Dana and I found ourselves juggling all those requests while also managing our own stress about our work deadlines.

Complicating the situation, six months prior I had bought Ellie a large trampoline as a holiday gift, hoping to make our new rental house feel more like home. Unfortunately, it was not until after I assembled it that I realized we might not be allowed to have it. After we dug up our tenant contract and read the fine print, my fear was confirmed. So I disassembled the trampoline and left it taking up space in the garage. On this particular day, our friend Susan called to say she wanted the trampoline and could meet us with her car to pick it up. I was delighted to find a good home for it, yet with all the needs and demands coming at us from every direction, it seemed very possible that neither Dana nor I would get any work done. I felt growing frustration, and it spilled out when I told Dana I felt we really needed to say no to something. She felt angry too and was frustrated that I lost my temper and that she also was not getting done what she needed to do.

The LORRAX process helped here, too. Although I was sure that trying to fit the pieces of the trampoline in the car so I could go meet Susan would just be a frustrating waste of time, we both *listened* to each other as best we could. Then I *opened* my mind to the possibility that the trampoline might fit. To my great surprise, it nestled right in. Upon *reflecting*, I realized that the unexpected redirection gave us the opportunity to solve multiple needs at the same time, and I agreed to meet up with Susan at a transfer point near my parents' house. I *released* my expectations around how much professional work I would accomplish, and I *acted* upon the plan. As it turns out, within two hours I had addressed all of the items that needed addressing, and I found myself quietly getting my work done at my parents' house while Ellie played with her grandparents.

Not everybody has the same resources available. I don't suggest that going to the parents' house is the best solution for babysitting issues, although in availing myself of that option, I'm probably in good company! The point is that using the LORRAX process and allowing ourselves to align with the circumstances—noticing resistance to the situation, being aware of my feelings of defensiveness with Dana, yet remaining open and reflective—allowed us to find a solution in which each person got what they needed. The value of seeing synchronicity all around is that even when circumstances seem not to be in our favor, we can find a way to improve our situation by living in flow.

Everything that happens—whether it involves our friends or our partners or our children arguing with us—is part of flow. Seeing our interactions with other people as part of the flow of life's events can help us cultivate vibrant and spacious intimate relationships.

Finding Friends with Flow

There is yet another aspect to relationships and flow that can be more uncomfortable for some of us. Flow doesn't just guide us to be more receptive within our chosen relationships; it also guides us to be more receptive to new relationships. If we pay attention to synchronicity, we never know who we might meet. The person sitting next to us on the plane, the individual in the waiting room auditioning for the same part we are, the colleague in another department whom our boss asked us to collaborate with—what positive experiences may flow from these connections if we notice them and act upon them?

One of my close friends would not have become my friend if I had not made an effort to align with the flow. Carl was an engineer on a team for which I was the project manager. I initially had negative preconceptions about Carl because he reminded me of somebody else I had worked with who put on a tough-guy persona. The previous person had often acted like a bully on our team, and I was not looking forward to repeating the experience. As a result, on our first video call, I responded to Carl's body language and countenance by immediately jumping to conclusions. He didn't say much, kept his arms crossed, and had an unreadable expression on his face. What's more, his credentials were impeccable, so he triggered my own insecurities. I imagined a future where he would question my skills or authority and I would never find my stride with the team.

I recognized this dynamic as it was happening, so I reminded myself to practice the LORRAX process—listening, opening, reflecting, and releasing before acting—and a shift gradually happened. It was not a single event but a progressive turning of the tide. For instance, at the end of the first call I expected Carl to be sullen and disinterested. Instead, he surprised me by saying he was genuinely excited about the project and looked forward to working with all of us. It would have been easy for me to miss this comment or dismiss it, but instead I *listened* and *opened* to the possibility that my previous interpretation might be faulty. Then I *reflected* on what he had said throughout the call and checked it for consistency with the possibility that maybe he was actually an amiable person. Upon realizing that I had at least as much evidence that he was likable as for the opposite conclusion, I *released* my attachment to my previous conclusion and tried to get to a sense of beginner's mind. My *action* was simply to continue to give him the benefit of the doubt on our next call and follow this LORRAX cycle again.

As I continued to listen openly to his input over the next few meetings, we navigated away from the valley of negativity we could have fallen into and toward a strong friendship. We eventually learned that we shared many interests and values in common, and Carl and his family have become close friends with me and my family.

My expectation was that Carl and I would not get along, but flow had brought us together on the project, and I was able to find the hidden gem in the situation and shape my world differently than I had expected. Although being open and authentic with people we already know and trust may be relatively easy, it is harder to be that way with people we don't know. Living in flow and synchronicity can lead us to new connections that we ultimately come to treasure. Maybe we can get help with a problem we are working on from someone we hadn't thought to ask, perhaps because we are a little scared to ask them for help. By choosing to step into flow, we can find important new connections that we might have avoided otherwise.

Our attachment to known relationships is so fundamental that we hardly notice it. "Team identification," as sports psychologists refer to it, has many benefits for social health; but it also has drawbacks that can be limiting.[57] In what Laloux[58] calls the "amber" stage of development, the beginning of civilization brought about a fundamental dependence on social identification for our survival, a trend also documented by Jeremy Lent in *The Patterning Instinct*.[59] Modern civilization depends on our general desire to be part of a team. This legacy exists today, whether we are talking about our political affiliation, our sports affiliation, or our association with our department at work. The effect can be benign or toxic; participation in organized sports can provide a sense of belonging that enhances personal well-being[60] and helps protect against suicide,[61] but it can also be associated with bullying[62] and racism.[63] This can be seen in politics as well. Identification with our "comfortable" relationships has benefits, but it also limits us to what we already know or believe.

The physics of flow suggests a new paradigm of relationship. When I need to solve a problem or I am seeking a meaningful experience, instead of relying solely on known relationships, I can rely on flow. Through the LORRAX process, unexpected events can arise that lead me to meaningful connections.

Although I may feel butterflies in my stomach when I need to break outside my comfort zone, unexpected matches may lead to the fertilization of rich new ideas. For example, when

> ### Finding My Thesis Advisor
>
> When I started my graduate program, there wasn't anyone in my department doing research I was interested in. In my second semester, I was assigned a class with a new teacher whose research looked interesting. On the first day, I went to the office to find the room assignments, and I glanced up just in time to see him walk out the door. I caught up to him to introduce myself, and I learned he was also lost, so we found the classroom together. I was the first student he met, and I became one of only two students he took on during his first year.

I was writing a section of this book that discusses the issue of racism (which you'll read in chapter 5, "Living from the Heart"), I wanted to get feedback to see if I had handled the topic sensitively. I sent the piece to a friend whom I trusted, but I also sent it to two other people whom I didn't know as well because I sensed they might have a useful perspective on the topic. Sure enough, the person I felt more comfortable with didn't reply for two weeks and was too busy to help. But in the meantime I received very helpful feedback from the other two folks, and in the process we built rapport.

Using flow, we can fluidly navigate a variety of different relationships, each of which serves us in unique ways. The importance of traditional structures such as family does not diminish when we live from flow. I am not suggesting we enter a state of relativism where we devalue the importance of close relationships or that we treat human intimacy casually. Rather, we recognize the commonalities we have with *any* person we meet, expanding our awareness to learn something useful from every connection that flows into our lives. We recognize that, in a responsive cosmos, anyone can bring us new opportunities or perspectives.

Is there really a risk in talking to this person with different-colored skin? Am I certain that I don't have anything in common with the woman or man sitting across the desk from me because they are much older or younger than I am, or because they dress differently from me? We can instead wonder if there may be some hidden synchronicity waiting to unfold with this person. We may share unexpected similarities we don't know about. Undoubtedly there are branches of the tree on which we do something great together. If we want to shape our world to find those branches, we may have to take the risk of connecting.

What can we do to uncover these hidden connections? We can follow the flow and start a casual conversation about whatever is natural to talk about. Then we can follow the LORRAX process and listen carefully for any clues of interest. While our upbringing may have encouraged us to form surface judgments about who we think we will synergize with, flow can allow us to gradually dissolve them by developing a new sense of trust. We shouldn't trust everyone and everything, but we can get into flow and see who shows up.

Flow in Organizations

It is one thing to incorporate more synchronicity and flow into our personal lives. This can be done relatively privately without affecting others very much. But what if we could bring this mind-set into our professional lives as well? I envision the impact as transformative.

During my tenure with a software company, I was promoted to the role of software project coordinator during a transition of leadership. The transition had the potential to be a positive experience for our team, but it was also a time of great uncertainty. Because I was now in a position of leadership, I was looking for an experience or project that would help the team bond in its new form. In my heightened state, my thoughts, feelings, and emotions were all trained on the urge for this situation to turn out well. I was naturally anticipating qualitative experiences—apples on the tree of possibilities—which helped the team and me negotiate this transition.

Within two weeks, a rare opportunity emerged to meet that need. The vice president of sales approached me with an unusual project and asked if I thought the team would be willing to take it on. It involved launching a small site for a new client within a two-week window, rather than the usual launch cycle of twelve weeks or longer. However, the client was huge and would be recognized by everybody in the department. For our small, quirky team it would be like meeting a rock star. Maybe this was one of those apples that would help our team bond.

I had been tapped for my leadership role partially because I tended to be realistic about what was achievable, so my instinct was to say no. But seeing the situation within the context of the LORRAX cycle, I decided to consider it more carefully and take it to the team. In talking it over, we realized that the project could be launched without any customization so that the entire two weeks could be devoted to configuration and testing. The team dove into the project with collective urgency, and when the project was delivered on time, a bonus was provided not just to those who worked directly on the project but to the whole department. We now had our team bonding experience! The timing of this rare opportunity aligned so well with our team's need that it seemed like a prime example of meaningful coincidence. By following the LORRAX process I was able to navigate my leadership role and help shape the situation for the best.

By and large, we do our life's work within organizations of one kind or another. Organizations are where the rubber meets the road and our ideas either fly or fall, based on their synergy with the ideas of others. We might be an educator, a graphic designer, a peace officer, a lawyer, a writer, or a stay-at-home parent; regardless, our interactions with others are where we ultimately make our mark on the world. Much work is already being done on incorporating flow and (to a lesser extent) synchronicity into the workplace. I am going to discuss three ways in which living in flow can influence the way we do business and run our professional lives.

The first way synchronicity and flow can positively influence us at work is to help us in navigating issues of control. Csikszentmihalyi's assessment of control in flow from chapter 1 is again highly relevant here. Flow is not about having more control or letting go of control, but about "lacking the sense of worry about losing control."

In *Simple Habits for Complex Times,* Berger and Johnston provide tools for dealing with our increasingly complex work environments.[64] They suggest that the countless variables and rapid changes in today's markets make the playing field fundamentally unpredictable and uncontrollable. In response, the authors articulate an approach to organizational development called "adult development," which conceptualizes change management inside complex organizations as a process of guided evolution. Because business decisions are so interwoven and context dependent, each person exercises some control over decisions. There must be an allowance for deviation and variation so the system can stay responsive to complex circumstances.

Similarly, according to Laloux,[65] in self-managed organizations everyone is accountable because there is no fixed job position and everyone has a stake in the game. The distributed power structure, as evidenced in the formal "advice process" but also embedded elsewhere in self-managed organizations, means that each individual must be comfortable handling unexpected input. Invariably the input received on an issue will bring up potential concerns that an individual hasn't yet considered, and the LORRAX process (or something akin to it) can be useful in turning unexpected feedback into effective action.

> ### Just the People She Wanted to See!
>
> Once when my family and I were attending a camp together, I had a long discussion with the camp director about synchronicity. She had scheduled a strategic-planning open house for the next day so she could take advantage of input from the adults who were on site for the weekend. She had never tried this before. Shortly before the meeting was to begin, a car arrived at the camp, carrying four philanthropists who were regular donors to the camp and who lived in the nearest city, more than two hours away. They had gone to a wedding in the mountains and happened to be passing by on their way back home when they decided on a whim to visit the camp. They unknowingly arrived in perfect time to attend the strategic-planning session.

A second way in which synchronicity and flow can be implemented in organizations is through communication style and habits. Some trends in the workforce indicate that today's employees feel more empowered and engaged in their lives at work. While promises of future pensions—and thus financial security—have become less common, employees have found themselves more drawn to find meaning or purpose in what they do.[66] They want their experience at work to be a positive, authentic one.

In *Crucial Conversations,* the authors emphasize the importance of creating safety in dialogue through a "Pool of Shared Meaning."[67] One can create safety by listening to when someone in the conversation feels unsafe. Through opening and reflecting, one can step outside the content of the conversation to reestablish trust. From the perspective of flow, clues may naturally arise in the course of the conversation—whether through casual comments, body language, or other forms of behavior—that can help one know whether reestablishing trust is needed and how to do it. Rather than being content to just make it through to the end of a meeting, the experience of flow can help us make our conversations at work more meaningful and authentic.

In *Nonviolent Communication,* Marshall Rosenberg emphasizes the importance of authentic listening and reflecting the needs of another person back to them when in dialogue, to ensure that we understand what they've said.[68] By seeking to understand the authentic needs of others, we elicit human empathy and encourage healthy interdependence, as we do in flow.

By changing our relationship to control and to communication, we can develop new methods of navigating synchronicity and flow in our organizations. When we implement structures and methods that rely on synchronicity and flow, we can be more flexible and responsive to today's diverse forces of change.

What are some of these methods? The U-process described by Senge and colleagues encourages us to see, sense, let go, "presence," crystallize, prototype, and perform.[69] This is another way to follow the guidance of synchronicity and flow, similar to the LORRAX process. They suggest that by shifting how we see the world, understand relationships, and make commitments, we come into better alignment within our organization.

In *Synchronicity: The Inner Path of Leadership* and *Source: The Inner Path of Knowledge Creation,* Jaworski encourages us to tune into the singular moments when our influence is most felt.[70] He says, "True leadership is about creating a domain in which we continually learn and become more capable of participating in our unfolding future. A true leader thus sets the stage on which predictable miracles, synchronistic in nature, can—and do—occur."[71] He points to the bottom point in the *U*-shaped innovation process, developed by Scharmer and colleagues,[72] where information can come from unpredictable and sometimes untraceable sources. We may not even realize where we get our crucial ideas—they may come while jogging on the treadmill or overhearing an ad on TV at the airport—and the idea of the trapdoor at the bottom of the *U* reminds us to look for and value these unexpected sources of information. "Leadership," he continues, "is about creating, day by day, a domain in which we and those around us continually deepen our understanding of reality and are able to participate in shaping the future."

Berger and Johnston describe a tool called "safe-to-fail experiments" that purposely creates room for the unexpected.[73] In this method of innovation, a team is allowed to undertake an experimental project in which the guardrails—the degree to which the project can get off track without anyone trying to fix or control it—are clearly defined and communicated. The guardrails are left intentionally wide so people have freedom to explore. Safe-to-fail experimentation creates a structure for innovation because it allows a predefined amount of room for unexpected events—what I would call synchronicity—to show up.

The authors of *Crucial Conversations* provide another method for staying in flow during interactions at work. They provide tools for noticing when conversations have turned unproductive due to a loss of mutual trust.[74] One may notice telltale signs of feeling threatened through one's physical state—maybe feeling tongue-tied, eyes becoming dry, or stomach clenching—or through one's emotional state, such as feelings of fear, anxiety, or anger. Once these signs have been identified, the authors offer three tools for reestablishing trust before continuing with the content of the conversation. One tool is to apologize when appropriate, e.g., "I realize I forgot to take your input into account in my proposal, and I'm sorry." Another option is to make a contrast statement to ensure that others know you have their best interest at heart, such as, "I do feel like you've done fantastic work overall, and I don't mean to undermine your confidence in making decisions. Here's what I need to request moving forward...." Finally, they suggest identifying or creating a mutual purpose, e.g., "We both feel strongly that the quality of the final product is crucial, even if we have different opinions on how to achieve that. Let's start from there."

This process is in great alignment with flow, because it teaches us that we must be aware of and responsive to the "vibe" in the room rather than simply continuing with our agenda obliviously. What the situation calls for may not be exactly what we had in mind, but by getting into flow and noticing useful indicators, we are likely to find a way to accomplish our agenda while ensuring that others also feel respected in the process.

In his doctoral dissertation, titled "Synchronicity and Leadership," Philip Merry articulates a method for incorporating synchronicity into business decisions.[75] Starting with a question or need at hand, we are encouraged to watch for events that can fit within the meaning of our project and then to respond with appropriate action. We pay attention to answers or results obtained from our first action, and this may lead to a "small Wow!" moment, a feeling of circumstances aligning with us. If we repeat this process, we may find a growing sense of "Wow!" as the project develops. A gradual realization of the larger context of the project or the connectedness of its parts may unfold from this process, leading to a "Big Wow!" Merry identifies factors that facilitate this process, such as letting go of the past (release), watching for unexpected situations (listen), being aware of one's own thoughts and needs (reflect), and opening to meaning (open). These factors align quite well with the LORRAX model. He points out benefits that arise from living in flow and synchronicity: for the individual, focusing on both now and the bigger picture, developing courage to face change, and growing confidence and authenticity; for the business, obtaining unexpected results and resources, connecting to the big picture of the future, and saving decision time; and for the team, stopping politics, helping negotiations, growing sense of awe and cooperation, and inspiring creativity.

Finally, in self-managed organizations, we shift our perspective on employment and employees overall. We aim to elicit the best from people by honoring them as creative, thoughtful, trustworthy, accountable, and responsible while also fallible and unique. We release the pyramidal hierarchy and the power differentials that go with it, investing the power of decision-making in those with the most knowledge of and investment in the issue. The sense of ownership and creativity that arises is ideal for cultivating flow and noticing synchronicity in the workplace.

All of these approaches can utilize synchronicity and flow to shape our work environments intentionally and for the better, even if not predictably. Decisions made in flow not only have the potential to provide benefits to everyone involved with the business; they are also more likely to benefit the outside world. When everyone on a team contributes authentically to the shared pool of meaning, the decisions that result are likely to account for all sides of the human beings involved. Since we all care about more than just our organization's bottom line, this process allows important issues from the community to make their way into boardrooms and workspaces.

You Can't Have Flow without Problems

Do you wish your life was free of problems? If so, you are obviously not alone. Yet this desire is a major obstacle to flow. Inviting synchronicity is about aligning with circumstances *as they are,* which we wouldn't have to do if everything was already aligned with our own expectations. It is only by coming up against obstacles that we can get into flow.

I recently rented a moving truck for our family to use in relocating to another city. On the morning of moving day, the rental company called me to say my reserved truck wasn't available, but a smaller one was. This concerned me because I was convinced that I needed a large truck. Yet I stayed in the flow by adjusting my expectations. When I arrived to pick up the smaller truck, they told me that one wasn't available either. Now I was stuck.

I felt frustrated, angry, and a little scared that my moving day might be derailed. I was stuck on my picture of how it was all supposed to go, and I felt my luck slipping through my fingers. Keeping the picture of the branching tree of possibilities firmly in my mind, I held my tongue and refocused to try to see the situation with new eyes. What constellation of events was happening? Where might a solution be hiding?

Instead of the instinctive frustration and blame that is so easy to end up in, a spacious pause ensued, maybe sixty seconds of all of us sitting with the uncertainty of the situation. What are we going to do here? Into that spaciousness walked the yard manager from his duties outside. He overheard us talking and chimed in, "There's a twenty-six-foot truck that just came in that's available, but only for the day." Amazing! I had been *listening* for just that sort of information, and now I had to *open* my mind because it was a really big truck, and I had to adjust my schedule to be able to bring it back in time. I *reflected* on whether we could make it work, *released* my concern about whether I would be able to drive the truck safely and whether I would get it back on time, and signed on the dotted line. This LORRAX cycle unfolded over about five or so minutes from beginning to end. We got a big truck as we had originally wanted, and we received a discount because of the company's mistakes.

Rather than focusing on the importance of the details they missed—Why didn't the rental company have my truck ready when promised? Why didn't they call me when the truck wasn't available?—I can see it all as part of a larger web of circumstances in which I ultimately had my need met. We did end up needing the entire twenty-six feet of the truck, so the fact that the smaller truck wasn't available on the day of the move was a small problem that ended up resolving a bigger problem. Living in flow allowed me to work collaboratively with the employees without generating more problems. I can still hold the rental company accountable for not following through on their commitment. I may choose not to hire them again. But because my situation turned out rather well in the end, the nature of my feedback for them is about helping them improve rather than venting about how they ruined my day.

Let's step back from this example, though. I don't think synchronicity is just there to help us save money and get the right truck. In the bigger picture, life is about learning and growing, and synchronicity shows up to help. What business isn't fraught with difficulties as part of its daily existence? Businesses need obstacles to strengthen their foundation and lead to greater long-term growth. What person ever developed something new without overcoming adversity first? Obstacles can serve the purpose of increasing our capacity as human beings and galvanizing us to accomplish the tasks we deem most important. Thomas Kuhn wrote about this in his analysis of what motivates scientists to pursue their research: "The man who succeeds proves himself an expert puzzle-solver, and the challenge of the puzzle is an important part of what usually drives him on."[76] If obstacles in science can be reframed as puzzles, can this same mentality apply to the flow of our daily life?

> ### Flowing from the Skies
>
> "I was driving to a meeting in a downpour when I realized I had left my umbrella at home. Unfortunately I was going have to walk a few blocks to get to my meeting from the parking lot, and I was going to get soaked. Just as I had that thought, a pedestrian crossed in front of my car with a First Republic Bank umbrella. I had recently opened a checking account there and had received the same umbrella as a promotion, and seeing the guy carrying the umbrella reminded me that my wife had left that umbrella in the trunk for a 'rainy day.' When I parked, I checked in the trunk, and there it was! I ended up staying dry that day." (Story contributed by George Scott)

Flow is about confronting appropriate challenges. Think about Csikszentmihalyi's ideal matching of challenge and skill, the flow channel described in chapter 1. He once performed an interesting experiment in which he tracked when flow states occurred throughout a normal day and how they felt to those who had them. He equipped participants with a pager programmed to send random notifications. Whenever they received a notification, they wrote down various aspects of their mood and what they were doing at the time. His results were somewhat counterintuitive, yet very familiar: we get high self-esteem from doing difficult tasks, but we don't enjoy doing them. He calls this the "paradox of work": "On the job, people tend to use their mind and body to the fullest, and consequently feel that what they do is important, and feel good about themselves while doing it. Yet their motivation is worse than when they are at home, and so is the quality of their moods."[77]

This is not only true of our workplaces as adults. He found that teenagers have the same experience with schoolwork:

> *Whenever adolescents are doing something they label as work, they typically say that what they do is important for their future, requires high concentration, and induces high self-esteem. Yet they are also less happy and motivated than average when what they do is like work. On the other hand, when they are doing something they label as play, they see it as having low importance and requiring little concentration, but they are happy and motivated.*[78]

This is no surprise to me as the parent of an eight-year old. If given her choice, Ellie will almost always choose to watch a computer screen or read a graphic novel, but she experiences lethargy and other symptoms of lower self-esteem when doing so. If instead we push her to do a difficult art project, practice her violin, read a book without pictures, or even clean the whole kitchen by herself, she resists fiercely, but once she starts we can't pull her away. When I come into the kitchen an hour later, the food has been reorganized in the cabinets and the kitchen table has been set, complete with name tags.

Obstacles are the hurdles that get us into flow. If we want to live more in flow, we are wise to choose self-esteem as our goal rather than ease and enjoyment. In the moment, self-esteem is correlated to flow states, whereas happiness is not. Yet the greater self-esteem we experience after the fact makes our lives richer and arguably makes us happier people. So the paradox of happiness is that if we seek it directly, we are unlikely to do the work necessary to overcome obstacles and get into flow, and the enjoyment we find will be fleeting. If we seek instead to act purposefully, we will face the obstacles necessary for getting into flow, and both our self-esteem and long-term happiness will be elevated. Instead of trying to escape from life's troubles, we are better served by asking how we can align with circumstances and do what is needed.

Yet the word "flow" is dynamic. It doesn't mean the same thing in every circumstance. Flow doesn't mean taking on every challenge. The joy of flow comes from finding what is right for you in the moment. To get into flow we need the *right* circumstances. According to research by psychologist Anne Wells, among women she surveyed, those who worked outside the home the most had the lowest self-esteem. However, all the women she surveyed *enjoyed* working outside the home. There are potentially a number of causes of these findings, but Csikszentmihalyi's interpretation is useful for understanding flow: "Full-time, professional women with families might have lower self-esteem not because they are accomplishing less, but because they expect more from themselves than they can possibly deliver."[79]

Self-esteem doesn't flow from taking the easy route or the hard one, but from taking the one that is appropriate to our abilities and needs in the situation. Getting into the flow channel involves being given appropriate challenges and then overcoming them. This is where I think meaningful history selection comes in. The concept suggests that we experience obstacles that will specifically help us grow in useful ways. Let's say Terry's goal is to be more successful in her role as CEO, and doing so requires her to become more communicative with her team around finances. Of the many branches of the tree that are available, let's say one of them involves something going wrong with her mortgage at home, a totally unrelated situation. Yet the problem with her mortgage engenders a communication crisis with her spouse, and the experience causes her to become more sensitive to the importance and power of clear communication about financial matters. This mortgage crisis was a synchronicity. According to the model of meaningful history selection, if Terry's mortgage crisis would end up making her more successful as a CEO—a change she was actively seeking—then her mortgage crisis would be more likely to happen. Her need to grow at work is drawing her toward outcomes where she becomes a better CEO, and the branch with the home mortgage problem is one of those.

If we set an intention, we should expect situations to arise that help us along the necessary path, whether we see how or not. Thus, it can be useful to approach obstacles with an open mind: "How might this frustrating, annoying, horrible challenge be helping me grow in ways I have intended for myself?" This is the reflection part of the LORRAX process. These obstacles are branches on the tree that, counterintuitively, lead to more apples, because they help us learn things that make us more likely to accomplish our apple-goals in the future.

If we want to find flow, obstacles become useful. Even though one branch of the tree of possibilities may be blocked, there may be other branches nearby if we look carefully. We can encounter obstacles with a light heart because we trust the tree. In our mind's eye we see not just the one branch in front of us but a multitude of possible branches.

Living in flow is more than just knowing about the concept of flow and experiencing it once in a while. As we become more familiar with what it feels like to align with our circumstances in a proactive way, we can stay in the flow more easily and have our actions build upon themselves. The next chapter takes another step toward integrating flow into everyday activities and lifelong goals.

4

BUILDING SYMBOLIC MOMENTUM

In mid-October 2017, actor Alyssa Milano made the following post on Twitter: "If you've been sexually harassed or assaulted write 'me too' as a reply to this tweet." Milano's attempt to draw attention to the embedded culture of misogyny that women experience in American culture drew hundreds of thousands of replies in just a few days. The "#metoo" movement—women telling the gut-wrenching truth about their mistreatment by men—had begun. The cultural tide led to the resignation of many powerful men from positions of influence.

But was that the true beginning of the #metoo movement? An activist named Tarana Burke used the phrase in 2006 as part of her efforts to empower women of color. Women have been talking for a very long time (mostly with other women) about their shared experiences of discrimination in the workplace. In 1991 law professor Anita Hill testified during U.S. Supreme Court nominee Clarence Thomas's confirmation hearing that Thomas had sexually harassed her, providing a momentary glimpse into this building undercurrent. More recently, Fox News's Eric Bolling and Roger Ailes were accused of sexual harassment in high-profile stories that made the main news cycles. During the 2016 U.S. presidential campaign, sexual harassment allegations were made against candidate Donald Trump, and the hashtag #NastyWoman was invented based on Trump's comment about opposing candidate Hillary Clinton in one of the debates.

What made October 2017 different? Momentum. Momentum of frustration and outrage had been building for decades or even centuries. It reached a tipping point in October 2017 and finally breached the surface.

What Is Symbolic Momentum?

The momentum evident in the #metoo movement is analogous to the concept of momentum in physics. An object has momentum when it travels on a certain path that tends to persist even after we change the conditions that caused it. What we call Newton's first law of motion can be stated thus: objects in motion tend to stay in motion, and objects at rest tend to stay at rest. The tendency to persist on a path also shows up metaphorically in daily life. If you've ever been on an eating binge, you've experienced this kind of momentum! Once you establish a pattern of going for one more cookie or another bite of turkey, your momentum makes it hard to stop.

We often think about momentum as the difficulty in slowing something down, like catching a flying baseball or trying to stop oneself from taking another trip to the fridge. But there is another side to momentum. Sometimes a system can be changing—gaining momentum—yet we don't recognize it until "all of a sudden" a major change appears. We could say that a fertilized seed underground has momentum in a certain sense, even though it cannot be seen. All the proper pieces are in place: if left alone under the same conditions, it will suddenly break out of the soil. Similarly, while we are bingeing, there may be a small voice that is gaining momentum below our conscious mind, a voice of self-respect and determination that suddenly breaks above the surface and says

Enough! We stomp to the refrigerator and passionately move all tempting items into the garbage can. This is the visible emergence of the momentum that had been secretly building.

These examples illustrate a process that unfolds via the usual physical laws of cause and effect: the plant breaks out of the seed because the right physical conditions of air, soil, and water exist. In contrast, meaningful history selection implies a "symbolic momentum" that builds as we get closer to apples on the tree of possibilities, i.e., "symbolic events." Remember that symbolic events are events that have a certain *qualitative* result, regardless of the physical setup of the world. For instance, satiating one's appetite is a *symbolic* outcome that can happen in many different physical ways: one can eat a carrot, a nutrition bar, or a steak. Building symbolic momentum toward the goal of satiating one's appetite would mean getting closer to any branches that include access to carrots, nutrition bars, or steaks.

When we move from point *A* to point *B* in figure 5, we find ourselves surrounded by more apples than before. At point *B* we no longer have access to the branch on the right, but since it had only a few apples, losing that branch has increased our chances of getting to an apple. We are now more likely to achieve our qualitative goal. Although we may not have been consciously aware of our progress toward the apples, we have moved closer to them—built momentum toward our goal—by taking this step.

We are always building symbolic momentum toward something. If we continue with the same choices—if we keep watering the seed every day—we will eventually find ourselves on a region of the tree that is full of apples. One of the apples is quite likely to occur. For instance, sending out a résumé to company *X* not only increases our chances that company *X* will directly offer us a job; it also puts us on a branch that is closer to more apples that reflect the *generic qualitative experience* of receiving a job offer.

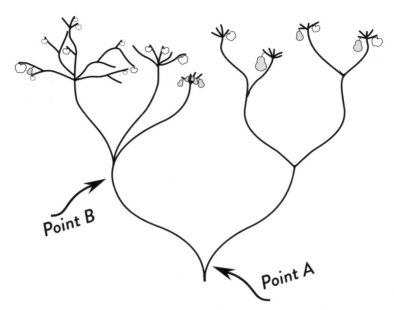

FIGURE 5. By moving from point A to point B, we move toward an area of the tree with more apples. We may not experience any evidence of getting closer to our goal, but we have built symbolic momentum and are more likely to reach our goal.

This qualitative experience can be broken down into actual mundane experiences such as the feeling of relief when depositing our first paycheck in the bank, or the pride of telling our spouse we got the job. If we keep on sending out résumés, setting up interviews, and networking, according to the theory of meaningful history selection we are building up symbolic momentum toward getting a job offer—or, more precisely, having something occur that provides that qualitative experience—from *any* place. We become surrounded by more and more apples, and eventually a job offer jumps out at us. The job offer may not come from somebody we actually talked to—it may come from a totally unrelated direction—yet it has a high likelihood of occurring because of our actions. This is symbolic momentum, and it works on our behalf in addition to the likelihood that each company we actually send our résumé to might, themselves, offer us a job.

> ### Wow, How Did You Get THAT Score?
>
> "I was taking a standardized test for getting into graduate school. During a break between test sections, I read a few paragraphs of a book I had brought with me for fun. When the test started again, the first question was an essay response based on one of the three paragraphs I had just read out of the book I had brought." (Story contributed by Orion Letizi)

Consider how I first found the spiritual community I play music for. I had been wanting to find a meaningful place to play music for over a year, and as a New Year's resolution I decided to start attending the Santa Rosa Center for Spiritual Living in hopes of breaking in to the music community there. However, after four consecutive weeks of attending services, it didn't seem to be working. The music director showed no interest in my offers of volunteer help, so on the fifth week I nearly gave up and stayed home.

I did end up deciding to attend service that day, and I was still unsuccessful. Yet when I arrived home I found a phone message from a woman named Reverend Mary, who was a minister for a different Center for Spiritual Living. She was inviting me to apply to be her center's music director. There did not seem to be any direct connection between my efforts to join the first center and her invitation to join the second center. I suppose it's possible there was an indirect connection, but because nobody really knew I was looking into this, and she hadn't been in touch with the first center, it seemed more likely that Mary's invitation was a case of symbolic momentum and synchronicity. Luckily, I recognized this event as a fulfillment of something I had been working toward—I listened, opened, and reflected—so I accepted the offer.

Because I had gradually surrounded myself by apples—anticipating the experience of being musically involved in a spiritual community, and acting accordingly—meaningful history selection suggests that something like that was bound to happen. Mary's invitation to me may *seem* spontaneous, but its momentum had been secretly building for a few weeks. This initial synchronicity was the gateway to an entire career path that has opened for me since, including meeting some of my closest friends.

Every action is meaningful in some way, so we are always building momentum toward something. If we aren't purposeful in our actions, we might constantly change our direction. For instance, if I went to the Santa Rosa Center for Spiritual Living three times, but then decided I *really* wanted to focus on developing my rock band to play at clubs and stopped attending the center, it would be like heading toward the apples in figure 6 but then switching to the pears. Merely building momentum doesn't benefit us; we need to actually reach the next apple to get the benefit. If we get *close* to an apple but then switch directions, we move to point C, where there are pears accessible but no apples. The apples are ripe but remain unpicked.

The timing of the call from Reverend Mary is an important part of what makes this a strong synchronicity. Think about it from my perspective: after a few weeks of committing to a certain path, I received a spontaneous offer of a paid job in a field that I happened to be looking to volunteer in. In fact, the offer came on

Building Symbolic Momentum

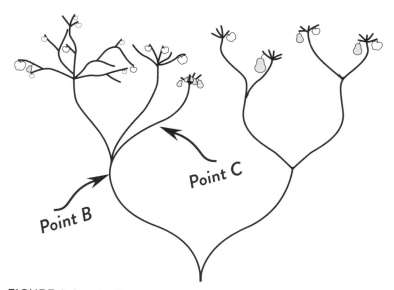

FIGURE 6. At point *B* we have increased symbolic momentum toward the apples, but if we change direction, we end up at point *C*, headed toward the pears. We don't pick the apples, even though we were very close to them a moment ago.

a specific day when I nearly gave up the quest but instead maintained my course.[80] This occurrence should not be construed as proof of synchronicity, but given the fact that I have never before or since been offered a job out of the blue—much less a dream job that aligned perfectly with my desired outcomes—thinking about it in terms of random chance seems like a poor explanation for the timing. Rather, the situation is consistent with the model of meaningful history selection.

Symbolic momentum can be bad news if we have been on a destructive cycle; it keeps us going back to the fridge for more turkey and pie. The apples we are heading for in this case might involve indigestion. Walking away from the fridge is not enough, because our momentum may lead us back to the fridge again before we know it. To change our momentum, we need to take bold action, like throwing away the leftovers and making it difficult to find anything to binge on.

But the momentum of symbolic meaning can also be a good thing if we have been consciously building toward an intended apple, as in the case where I became music director for Reverend Mary. In these cases, momentum can help carry us forward through moments of doubt.

Boldness Shifts the Probabilities

So why don't we experience meaningful coincidences more often? As discussed in chapter 3, catching synchronicity or getting into flow often requires a spark from us. To shape our world, we need to be able to see the potential latent in a situation and do something to actualize it. Even when meaningful or desirable possibilities exist, they may require our participation in order to unfold. We are the spark that makes synchronicity happen.

Remember that the first step in meaningful history selection is to define what the apples are. Assuming we have first examined our hidden feelings so we are creating apples of personal breakthrough instead of self-sabotage, the goal is to reach a point on the tree with an imbalance of types of apples. We want to set up an abundance of apples on one path and very few on the other. This stacks the deck toward a meaningful outcome or synchronicity.

How do we provide the spark to create a tree whose apples are unbalanced? If we take an action that leads to responses from the world that are very different from each other, one of those responses will align strongly with our action, and the others less so. That one branch will align better because it will involve some external event that leads to many more apples. In other words, by acting boldly in some direction we give the cosmos the chance to respond in a way that strongly supports our intention.

Boldness is another way of saying that we bring an authentic emotion into our action and maybe take a risk. Under the ideal circumstances, we bring emotions of love, joy, and excitement, and this amplifies our pull on the constructive experiences we wish to create. Of course, the opposite is possible too. I find myself very bold when I am angry, but the outcome isn't usually constructive. When I'm angry at my daughter for carelessly spilling her drink or at a vending machine for malfunctioning, I seem to be willing to do and say things I would never do otherwise. This form of boldness is very powerful, but it is often directed toward outcomes that aren't aligned with my highest good. Whether our bold action comes from love or fear, our clear and strong emotions pull experiences toward us that reflect the emotion.

Synchronicity Makes the Grade

In the final ten minutes of studying before my Statistical Mechanics midterm, I overheard a friend say, "Our teacher told us to study the two-state magnetization problem. You'd better do that." The comment wasn't directed at me, but I followed the advice anyway, and the problem indeed ended up being right there on the exam. While working my way through the test, I went for maximum points as quickly as possible, which meant saving the first problem for last because it looked nasty. With thirty minutes left, the teacher realized that the first problem (which I hadn't started) was too hard, and he removed it from the exam. I had optimized my time perfectly, and I killed that exam.

Why is boldness necessary? Shaping our world means bringing about events that are different from those we are currently experiencing. Acting boldly shifts away from the default course of events into an area of the tree more aligned with our intentions. Such a purposeful shift requires investing energy, because in the meaningful history selection model we are increasing order in the branches of the tree of possibilities.

For instance, raising our hand in a meeting to suggest a new idea or to counter the accepted viewpoint may be necessary if we want to avoid just going down the default path. On the default path, complexity, vibrancy, and order tend to decrease over time. Think of what happens to a company that doesn't periodically renew its vision and strategy: the company gradually deepens its ruts of behavior and becomes less innovative, less adaptive. By contrast, the result of acting boldly is to purposefully create more order and new patterns, shaping the layout of the apples on the tree in a new way.

On the tree of possibilities we can visualize this increase or decrease in order. Let's compare the trees in figures 7 and 8. In figure 7, the branches have equal numbers of apples, so neither branch increases our odds of eventually hitting an apple. This is a nonmeaningful layout of apples. Although it may seem very ordered because everything is quite uniform, from a physicist's point of view this layout has a minimal amount of order, because all the branches are identical. To understand why physicists consider uniformity boring, imagine if all the atoms in the universe were spread evenly throughout space. It would be a uniform cosmic soup, and we would say it had no structure or order. There would be no molecules like sugar or cellulose to create life, not to mention planets and stars.

In figure 8, if we end up on the left branch, the likelihood of getting to an apple is higher than if we end up on the right branch.

Building Symbolic Momentum

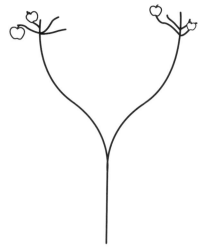

FIGURE 7. This tree has equal numbers of twigs on each branch and equal numbers of apples on the twigs. This is a nonmeaningful arrangement.

FIGURE 8. This tree has different numbers of twigs on the left and right branches, and it has different numbers of apples spread across each branch. This is a meaningful arrangement, because by going left we have a different chance of hitting an apple than if we go right.

This is a meaningful layout of the apples that has more order than the layout in figure 7. We can clearly distinguish between what happens if we go on the left branch and what happens if we go on the right branch. This is the heart of living boldly: we clearly distinguish between the path we take and the path we don't take.

To see how boldness changes our symbolic momentum, let's consider the trees in figures 9 and 10. If I have worked hard at my place of employment for six months straight, I will have gone from a tree layout that looks like figure 9 to a sub-branch of the tree that looks like figure 10. I start off having to act boldly in order to get to the left branch with more apples (positive work outcomes), but once I am there, the apples will be more or less evenly spread over all the branches. In other words, without too much further effort I will continue to have success at work because I am already on that path.

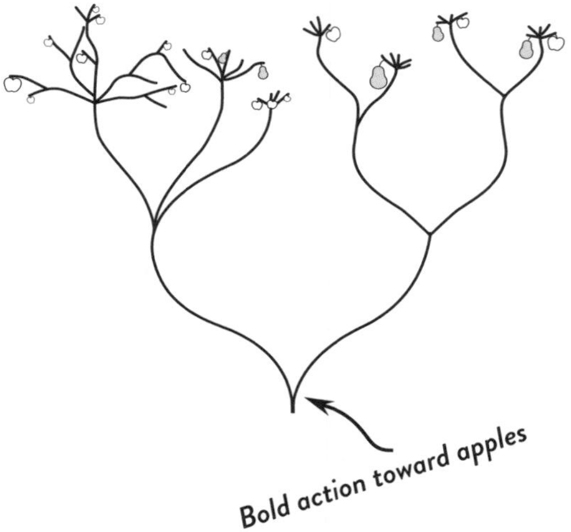

FIGURE 9. By taking bold action aimed at high achievement at work, I pull myself closer to the apples of positive work outcomes on the left branch.

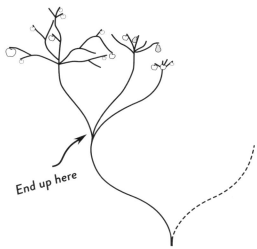

FIGURE 10. Later I find myself on the left branch, close to many apples (positive work outcomes), which are now evenly spread around the tree branches. However, I do not have a high chance of encountering positive family outcomes, the pears.

If, however, I want to have a different type of experience unfold, such as positive *family* outcomes (pears), those opportunities are not likely to unfold without some effort. I have put myself on an area of the tree where there are lots of apples but very few pears. If I do more of the same, which means climbing closer to apples, the chance of me landing on a pear is quite low. My work will thrive but my family may suffer. In order to enhance the chances of getting to the pears, as in figure 11, I have to do something bold that prioritizes my family life, like a spontaneous family movie night.

We started with apples bunched unevenly on the left in the tree, but then we went left, and within that branch the apples were fairly even. We built symbolic momentum toward apples. The figures show us that yesterday's bold action becomes today's default pattern. It initially took bold action to lead toward apples, but gradually that became the default outcome.

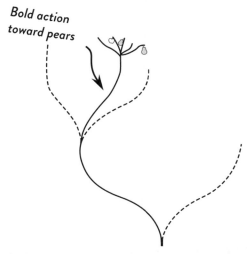

FIGURE 11. Only by a different sort of bold action, with positive family outcomes in mind, will I be likely to end up moving toward the branch with more pears.

Boldness is unique to the moment. It requires a dynamic, self-aware appreciation for what is needed in each situation. Jaworski says, "Commitment begins not with will, but with willingness."[81] This all comes together in the process of flow. In flow, our actions are not premeditated. We find a balance between listening carefully to our environment and acting boldly to shift the probabilities. From the illustrations above, though, we see that a passive acceptance of the way things are doesn't lead to constructive synchronicities or synergistic change. Flow involves both accepting circumstances as they are and looking for opportunities to act boldly and shift them.

Csikszentmihalyi also views flow in a proactive way, as the balance between challenge and skill. "For each person there are thousands of opportunities, challenges to expand ourselves," he says. "Such experiences are not necessarily pleasant at the time they occur."[82] He reminds us that the experiences that contribute most positively to our life flow, the ones that generate new patterns, are sometimes uncomfortable. By accepting—or even choosing—uncomfortable situations that serve a purpose, we shift our momentum on the tree and alter the shape of our world.

It's important to emphasize that nothing in this description is placing value on one branch or another. Bold actions separate out the wheat from the chaff, differentiating between the branches we choose and those we do not choose. The branches are not good or bad but rather just a reflection of our decisions. Meaningful history selection is a process which involves an initial action, the anticipated experiences behind the action, and the response from the cosmos. It doesn't bring us closer to what we *want* but rather closer to what we *choose* through our actions. The process is neutral.

Sometimes apples are important, and sometimes pears are. According to the theory of meaningful history selection, what we can count on is that we will tend to experience outcomes that, over time, reflect the nature of the choices we make. The opposite of a bold action—call it a "timid" action—is one that doesn't lead to different outcomes on the two branches. In the timid case, we don't really make a choice with our action, and the probabilities don't shift. Neither bold nor timid is better from the point of view of the underlying physical process; the terms simply describe the process of choosing and the meaningful consequences that result.

To understand how this plays out in a real example, consider Evita's experience of trying to get into her preferred graduate program, from chapter 1. The original situation is depicted in figure 12. The branches with apples are those on which she has the anticipated experience she is seeking: getting into the graduate program. There are only a handful of apples, and they are spread across all the branches of the tree. The chances of her getting onto one of those branches is not particularly high.

By taking the proactive, intentional action of calling the department to follow up, Evita defines for herself what the apples are, and those branches gain weight. Still, though, in figure 12 the weighted branches are all mixed up with the empty branches.

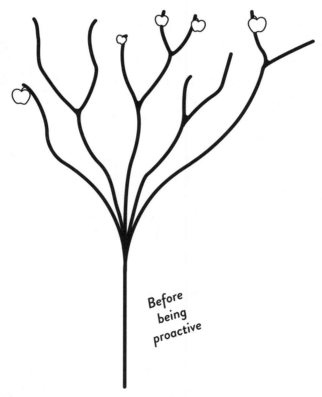

FIGURE 12. Before Evita is proactive in calling the school, the apples representing situations where she gets into the school aren't even defined yet, but we can imagine that those branches are dispersed randomly throughout the tree.

She hasn't really shifted the odds of the overall positive outcome. The crucial step is that any event that leads to future apples will now become more likely due to the weight of those apples. The synchronistic event S is one such event. S represents the case in which the head of the department is somebody she knows. This scenario would naturally increase her chances of reaching an apple any number of different ways. The head encourages the committee to take a second look at her application, and they realize a spot just opened up for someone with her qualifications. Another possibility is that a faculty member had recently put in a request for a graduate student with skills precisely matching Evita's background. The fact that Evita had a personal connection with the department is an event that leads to all these possibilities, and because there are so many apples branching off of this event S, the branch is very heavy. (See figure 13.) This particular event becomes more likely to happen, even though it was totally unexpected for Evita. From her perspective, it feels like a meaningful coincidence, a huge lucky break. From a wider perspective, an event occurred due to her actions that may not have occurred otherwise. This is what is meant by shaping her world.

By comparing figures 12 and 13, we see that Evita's proactive effort divided the existing branches into meaningful groups. The group on the left is much more likely to contain an outcome she was aiming for than the group on the right. The group on the left is all the branches associated with the potential fact that she knows the department head. Naturally, it contains more apples than the branch on which she has no personal connection, so this potential fact became more likely. The grouping is not perfect; there remain some apples in each group. But we call her action "meaningful" because there is a significant difference between the number of apples on the left branch and the number on the right branch.

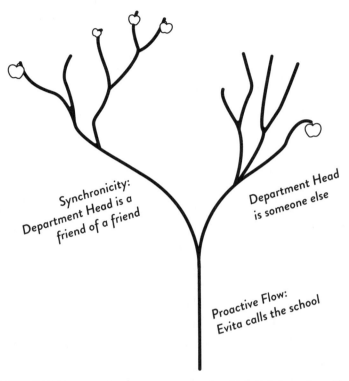

FIGURE 13. Because there are many ways in which Evita's goal could become a reality from the circumstance S in which she happens to know the department head, we can divide the branches into two groups and find that more of the apples are on the left group (S). Because the apples represent weight, the S branch becomes heavier; hence, S becomes more likely to happen. In order to obtain this situation, Evita needs to be proactive in calling the school, which clarifies objectively what her desired outcome is and defines the apples.

The tree diagram illustrates how we can build symbolic momentum toward certain types of outcomes because of meaningful history selection. Without this principle, we are truly at the whim of circumstance. We can build toward an outcome and have it snatched away from us at the last moment. While this is still true with meaningful history selection, the tree shows us that even when we work hard toward an apple and then miss it, there are other apples nearby. If we can recenter ourselves and look around with an open mind, using the LORRAX process, we might see unexpected opportunities to grow and thrive even at those times when we think we've missed our chance.

The Symbolic Momentum of Today's Crises

Symbolic momentum gives me hope about global situations like climate change. Most of us are aware that even if we completely changed our consumption habits tomorrow, the effects of the climate crisis would take time to ease. This is because both our human technology and the planet take time to adjust after more than a century of investment in carbon-based technologies. But according to the model of meaningful history selection, climate change is also a result of the symbolic momentum we have accumulated on the tree of possibilities.

When we choose to commute alone by car or purchase shrink-wrapped vegetables at the store, we keep bringing ourselves closer to apples on the tree that, frankly, aren't so good. We have shaped the world we currently have, in both its successes and its failures. Even if we try to personally unplug from the system by making low-impact choices, the various aspects of our economy are so interrelated that it is impossible for any one of us to avoid having a negative impact on some aspect of our world. Within the framework of meaningful history selection, it may be possible to understand facts such as extreme weather events as physical outcomes that are meaningfully related to the collective choices we are making.

We might also see random acts of violence or political gridlock as examples of symbolic momentum. Are they accurate reflections of our overdeveloped sense of self-reliance and our underdeveloped ability to connect with each other? By building this symbolic momentum, we have surrounded ourselves with these types of branches. One tragedy or another becomes inevitable.

If these crises are guided by the symbolic meanings of our actions, how exactly do we quantify what the symbolic meaning is? Here I will need to be a little vague even as I try to be precise, for I don't think there is only one right way to describe the symbolic meaning of today's problems.[83] The meaning of an action was defined in chapter 2 in terms of the qualitative experiences the action leads to. All the possible scenarios that end in a certain qualitative outcome get grouped together as, say, experience X. For instance, we can group the branches of the tree according to this question: "On which branches does the stability of our climate get better (X), and on which branches does it get worse $(\bar{X})$?" Then, when we do something that disregards the environment, our action aligns with many of the branches in $\bar{X}$ because these branches mirror the anticipated qualitative experience—a world in which our environment doesn't matter.

As a result, any event that leads to this final condition ("planet becoming less stable") becomes more likely. Suddenly, seemingly out of the blue, we have a massive oil spill on the Gulf Coast or a massive earthquake near a nuclear power plant in Japan. Now, these specific events clearly become more likely the more we drill for oil or use nuclear energy, but I suspect they also become more likely overall due to meaningful history selection, as the cosmos responds to our collective choices.

We don't have a good scientific repertoire of words to describe qualitative experiences precisely, so I have to be vague in describing the anticipated qualitative experience as "a world in which our environment doesn't matter." This is really shorthand for a whole collection of personal experiences one might have: the experience of missing the Super Bowl because you get sick from vegetables tainted by *E. coli,* or the experience of having brown tap water when your friends come over for dinner. These are just a couple of fairly benign examples of mundane personal experiences that might occur under those circumstances.

While we can look at each of our individual choices under this high-resolution microscope, we can also step back and look at the big picture. Many cultures acknowledge a fundamental balancing act between polarities, whether it is yin and yang, divine feminine and divine masculine, Shakti and Shiva, heart and head, and so on. The trend for many centuries, perhaps since the beginning of agriculture twelve thousand years ago,[84] has been an emphasis on what traditional Chinese medicine calls yang values (such as differentiating and competing) over yin values (such as connecting and collaborating).

Our modern technological civilization is largely built on this yang approach. The word "science" comes from the Latin word for knowledge, which is thought to be related to the Latin root *scindere,* which means "to cut, divide." Symbolically, we can imagine how climate change may reflect these same principles. Our suppressed emotions are rising up to redefine socially acceptable topics of conversation—think about the #metoo movement—and the ocean levels are simultaneously rising to overcome our frail and unprepared coastlines. Our brittle financial systems, our overwhelming reliance on a single type of fuel, and our concrete houses laid out in neat little rows are ultimately no match for the complexity and raw power of Mother Nature, or of human nature for that matter. Our atmosphere is trapping heat—the byproduct of the combustion we worship—giving us too much of a good thing.

This pattern can be seen in social issues as well. We use coercion instead of compassion to keep people in line. We focus on the differences between people more than their similarities, so that power differentials exist between women and men, majorities and minorities, children and adults. Even the decision to drive alone to work occurs within the context of a society that requires each member to contribute in exactly the same manner (employment) using exactly the same medium (money) in order to obtain fundamental security. We measure the contribution of every single individual down to the minute, and we compensate it accordingly, therefore often requiring people to make decisions that are detrimental to the planet simply because they are trying to make it through the day. This entire context arises from the yang perspective of separateness, the head dominating the heart.

It seems to me that all these factors are reflected in the emergence of climate change as a transformative force of our era. Climate change affects everybody, reminding us that our superficial distinctions among people are not as important as we make them out to be. I suspect that climate change will gradually bring us together in our humanness despite our differences, eliciting our innate compassion and desire for connection.

This is why understanding symbolic momentum makes me hopeful. When we rebalance the quality of our choices, for instance by living in flow, we build symbolic momentum in a different direction. When we live in flow, sometimes we push for our own ideas and sometimes we allow other people's ideas to rise to the top. Rather than maintaining control (yang) or surrendering (yin), we let go of the worry about maintaining control, and we flow between the two effortlessly. Rather than trying to forcefully shape our world according to our own concerns, we let our choices speak for themselves and allow flow to do the shaping.

Living in flow is an alternative to being strictly yang or wholly yin, but it is not a foreign feeling to most of us. It is perhaps the most intimately familiar feeling for those who had a reasonably healthy childhood; it is the experience of play. Combs and Holland say, "Play may be thought of as the feminine side of the masculine enterprise of exploration and discovery, while these enterprises are the masculine side of the feminine activity of play."[85] Here it is important to keep in mind the nongendered use of the terms "feminine" and "masculine"; they refer to a set of qualities that any person can have and that are not limited to one gender or the other. So when we are struggling with finding the flow, we can remember the feeling of playing. What turns us on? What gets us excited? Where is our attention naturally drawn?

As humanity makes more choices through play—acting from compassion and collaboration as well as competition and differentiation—synchronicities can happen that rebalance our circumstances both locally and globally. Even though the world may still contain some terrible news each day, we will see—we are already seeing—more positive synchronistic connections occurring as well. Not every event is salient, and life will always consist of both negative and positive events, so the effects of our efforts in this direction may not always be obvious. Yet by living in flow we can expect a shift away from an overall feeling of separateness and helplessness to a sense of belonging and empowerment.

Symbolic Momentum Can Change Overnight

Clearly, any prescription for effectively addressing these major human issues would be welcome news. But changes in worldview are not easy to bring about. So why does this approach make me hopeful? There are two reasons why I think the possibility of symbolic momentum would be good news for us. First, symbolic momentum can give us a form of faith to carry us through dark times, which we'll talk about at the end of this chapter. Second, we can change symbolic momentum very rapidly, which we will talk about here.

Changing the physical circumstances of climate change is a very slow, labor-intensive process. It requires passing laws, inventing and building new technology, and thousands of other little steps repeated everywhere throughout the world. Symbolic momentum, on the other hand, can change overnight. This happens any time an entire population changes its view of a situation based on one or two salient events. Symbolic momentum changed overnight in the "shots heard round the world" at the beginning of both the American Revolutionary War and World War I, and at the bombing of Pearl Harbor. Symbolic momentum changed the day the Russians launched Sputnik 1 into earth orbit, triggering an injection of money and enthusiasm into the U.S. space program. Symbolic momentum changed when Rosa Parks refused to give up her seat on a bus in Montgomery and when peaceful marchers were attacked in Selma, leading to major gains in the U.S. civil rights movement. Symbolic momentum changed on September 11, 2001, as millions of Americans suddenly felt for the first time that their country was not a safe haven from political danger. All of these were bold events that instantaneously reshaped the course of history, for better or worse. Changing our minds can happen overnight when the appropriate spark is applied.

Now picture a future moment in which we have already shifted our symbolic momentum concerning climate change. Our whole population is motivated to do what it takes to address the issue by recognizing the power of our individual actions and behaving collectively so as to aim for some good apples of balance and wholeness on the tree. It is within the realm of possibility that three months from now a breakthrough will occur in energy storage technology, cutting the energy wasted on the electrical grid by 90 percent and reducing our power needs by an order of magnitude. Such a technological breakthrough could be the result of a series of small synchronicities just waiting for us to put the weight on those branches.

Or imagine that an enormous extreme-weather event occurs shortly before an important election, resulting in a public backlash that votes out of office many leaders who deny the facts of climate change. The timing of such an event is within the possibilities of the responsive cosmos to orchestrate. Suddenly we may find ourselves in a political environment within which progress is possible.

Imagine that as we continue to build symbolic momentum toward wholeness and balance, a viral movement arises online that transcends government gridlock and sparks a worldwide commitment between individuals in every country to change their thought patterns and consumption habits. This, too, is simply the alignment of events to conspire in our favor. This is precisely how I suspect the responsive cosmos works.

But in shaping our world, we are always in charge of the first step. It is not our job to worry about how we are going to get to the finish line; our job is to build symbolic momentum in the direction we wish to go. Each step we take toward balance and wholeness brings us onto a branch of the tree of possibilities that has a slightly higher density of apples representing positive change. As the density of positive apples increases, I suspect we will experience more and more synchronistic events that allow us to respond proactively to further our mission. Those apples are there, somewhere on the tree. We just need to get to them.

The change we want to see comes from inside ourselves. To address the big issues, let's begin inside by shifting our workplace mentality toward joy and authentic expression, and away from just doing our job. Let's begin inside by shifting our cultural tendency (if we are men) to respect the contributions of men more than women, or (if we are women) to defer to men even when we have more expertise on a subject. Let's begin inside by acknowledging and respecting the contributions of all people in society and transforming our personal judgments into mutual understanding in every relationship we have. Let's begin inside by speaking up for ourselves and what we care about no matter who we are with. These are ways in which our choices can shape our world. Without such changes, we are unlikely to find the systemic solutions that are needed for issues as enormous as climate change or globalization.

Starting with inner change means focusing more on the behaviors in our own lives that bring about discord and make it impossible to find common ground with others in our community. The real problem that climate change presents is not that we are putting too much carbon into the air—a fact that is actually not difficult to remedy; it is that we are unable to talk to each other authentically about putting too much carbon into the air. The problem is not the *content* of the situation, but the *context*. Solutions abound for the issue of climate change—and for gun violence, trade globalization, and urban traffic, for that matter—but if we are unable to see the world through each other's eyes, then we can't get to those solutions on the tree.

It is easier to see through the eyes of another when living in flow, because we are constantly listening to, opening our minds about, reflecting deeply on, and releasing our own preconceptions of what others say and do. I envision global solutions arising not from individual leaders acting in a top-down way, but from a network of engaged, aware people who are working together in flow. They are working in local communities, keeping in mind the best interest of themselves and everyone else; and when those solutions work well, the network propagates them effortlessly to other communities.

When we live in flow, we can lead fulfilling lives and pursue happiness in synergy with each other. By adjusting what goes on inside ourselves, rather than trying to fix the outside, we may collectively solve global problems by participating in society in the ways we each feel drawn to contribute. As our personal lives become expressions of our authentic selves, we can't help but become a community of people capable of finding workable solutions to practical problems.

But if the first step in building symbolic momentum is always up to you, how do you take that step? How do you make a difference in whatever issue you care about?

Your Job Is to Leap

One time a man named Matt went on a river rafting trip, and his boat stopped for lunch in an area near some rocks. People were having fun jumping from the rocks into the water. Matt was hesitant to do it because it looked like there were jagged rocks below, and one could get sucked into a dangerous downstream area. When he mentioned this to the river guide, the guide said, "Your job is to leap, because once your feet hit the water the flow will carry you safely into the current and downstream."

Immediately Matt recognized what he was being called to do. Without hesitation he climbed up onto the rock and leaped into the water. And that wasn't all; he got out of the water and jumped back in over and over, caught up in the exhilaration of the flow.

Jumping off the rock is a metaphor for taking bold action in the world. We might feel held back by fear of the outcome, but we can trust that the flow of the river will take us around the rocks and that the flow of life will carry us through various circumstances. We may sometimes hold ourselves back because we are afraid there is no way we could be successful. But if our actions are always building symbolic momentum, then we are encouraged to keep working toward an experience we want to create, even when it feels impossible.

Meaningful history selection is like a mirror for what we believe. We make our decisions through a filter inside our head, and we interpret the response from the world through the same filter. Decisions made in this way don't take all the options into account; they only consider options visible through that filter. In the distant past, the Colorado River had many possible paths as it wound through what would later become the American Southwest. Yet eventually it reinforced a particular path until it had eroded a channel so deep that it became the Grand Canyon. Our beliefs become more deeply graven on our hearts over time, just like the Colorado River.

Even though many of our beliefs may have evolved due to this gradual reinforcement of certain experiences, we often hold our beliefs to be true so strongly that we take them for granted. We are prone to what psychologists call confirmation bias. Cognitive scientist Donald Hoffman has made the case that, from a natural selection standpoint, human perception did not evolve based on how accurately it mirrors actual circumstances but instead based on how well it confers survival fitness upon the organism.[86] So it's quite likely that many of our beliefs do *not* represent the way things really are, especially if those beliefs make us feel good.[87]

Meaningful history selection can play a role in this process. Our initial actions may be driven by unconscious feelings, which influence the weights of certain branches and lead to a particular meaningful experience in response. We take another action influenced by what we learned from the first experience, and this may shift the weights of the branches again. If our filter is accurate, that's great. Each successive action will then become more effective and better aligned with the circumstances.

But any errors present in our perception can become amplified. The amplification happens because the weights of the tree branches are influenced by our actions. For instance, when we misinterpret something a family member says to us, we don't tend to respond scientifically by controlling all the variables to discover the true nature of the situation. Rather, because we are strongly influenced by our feelings, we tend to take actions that reinforce the misinterpretation, creating stronger filters and deepening the banks of the river.

If we can let go of a fearful belief long enough to give us courage to leap, as Matt did, then we can reroute our momentum in a new direction. I had a memorable experience of this in the beginning of my freshman year of high school. I desperately wanted to make friends with girls, but I felt unattractive and awkward. Instead of socializing at lunch I went down to the arcade a few blocks away and lost myself in video games. By some miracle, after a few months I realized that this was not how I wanted my life to be at my new school, and I decided this was my chance to be different.

I began occasionally staying on campus for lunch and built momentum by taking little risks to interact and sit with people. In my teenage awkwardness, there were many moments each day where I could interpret things according to my old story—that nobody really liked me and I was an outcast—yet the positive feeling I got from some of the interactions was so compelling that I felt just barely confident enough to keep trying. I was holding onto an anticipated qualitative experience of "making friends" (which is a complex mix of emotional experiences that I won't try to untangle). I had just enough boldness to take advantage of some of the opportunities that came up, and I gradually shifted the banks of the river before they became too deep.

The effort paid off. Within a month I had not only my first girlfriend but also the beginnings of a more important friendship—with the girl who would eventually become my wife and soul mate. At that time, my beliefs about myself were highly malleable, and I have thought back many times to what would have happened if I hadn't shifted course. I might have a very different set of beliefs about myself today.

Rather than deciding *what* we believe, the deeper choice we each must make is *how* we interpret our experiences: from fear or from love. Fear limits the breadth of our view and constrains our options, whereas love keeps us open to alternate interpretations and helps us see things through the eyes of others. Psychologist Richard Wiseman performed an interesting study on luck, and he found that to a significant extent people create their own good and bad luck based on their interpretations of events in their lives.[88] When we act out of fear, we interpret the world's response within our preconceived notions, which inherently limits us. Our fear closes us down, reinforcing itself and maybe even building momentum toward what we are trying to avoid.

We will always be faced with the possibility of fearful thoughts in any situation we encounter. Thus, I think we need to learn to recognize fearful thoughts when they arise and not create reality from them. Our fearful thoughts are not *the* reality, but they can *become* the reality, especially with the help of meaningful history selection. If that happens, then we will probably influence the selection of self-fulfilling circumstances that provide evidence for our beliefs, leading us to become further entrenched in them.

I think it is our job to treat belief as a muscle, and like any muscle, belief needs exercise. This means exercising our ability to choose what we believe. I don't mean freedom to choose alternative facts. Nor do I mean just seeing the bright side of things with a naïve smile. I mean deciding which of our *own thoughts* we choose to believe. This decision has particularly important consequences, if meaningful history selection is correct, because

> ### Finding the Right Job
>
> "Each evening, after full work days at my perpetually struggling company, I sought my future career. I hungrily chased 'compromise opportunities' that were less in alignment with my career goals. After more than a year of exhaustive searching, I changed my thinking from 'picking the least bad option' to imagining the most positive possibilities. I arranged a presentation of my portfolio in the rarified field of electric motor design, and I posted to social media. After several weeks with no response, suddenly I received three cold calls from three genuine contenders and converted each into on-site interviews in the space of a single week! The third interview—and the least appealing to me, due to its distance away and its massive and contentious global presence—turned out to be the best and only job offered me. It was the opportunity most likely to succeed in bringing forth my gift and manifesting my intentions to make a positive impact. I gladly accepted the rewarding opportunity and responsibility to develop high-quality electric motors in vast quantities to replace fossil-fuel transport, beyond what I had envisioned for myself. My change in mind-set opened me up to calling in the right opportunity." (Story contributed by Edgar)

the beliefs we act upon influence the types of experiences that flow into our lives.

The *Tao Te Ching* says:[89]

*In the pursuit of knowledge,
every day something is added.
In the practice of the Tao,*

every day something is dropped.
Less and less do you need to force things,
until finally you arrive at non-action.
When nothing is done,
nothing is left undone.

By letting go of preconceived beliefs we loosen the cycle of entrenchment, and we feel more free to leap into life when the right opportunity arises.

Faith in Flow

Recognizing the symbolic momentum in our lives—in other words, staying alert to what sorts of circumstances we are consciously building toward—can have a big impact on how we perceive the world. This kind of awareness can make us feel immersed in an ocean of meaning. It can allow us to have confidence that we are on our path, even if it doesn't feel like it in the present circumstances. For example, when looking for a job most of the actions we take don't directly lead to landing a good job, yet we can rest assured that investing time in these "unsuccessful" activities builds momentum toward the opportunities that will likely come our way from some unrelated direction if we persevere.

Confidence in the unseen is what I would call "faith." But faith doesn't have to be blind. Symbolic momentum and synchronicity can give us a form of faith that is both explainable and testable. If meaningful history selection is correct, I suspect it encourages us to have faith that the experiences we anticipate and work toward will ultimately manifest, although in an unpredictable manner.

There are some similarities between this definition of faith and the faith of religious belief, and I wonder if these commonalities can provide a bridge between the scientific and spiritual worldviews. For instance, if we consider prayer as an inner state of being in which we anticipate future qualitative experiences or conditions, then meaningful history selection appears to be compatible with sincere religious practices.

This would imply that prayer of this nature is often unconscious. In this view, we are always praying, because we are always anticipating the qualitative experiences we are drawn to or are averse to. One who sincerely practices prayer is making this process conscious and seeking to direct it toward specific outcomes. But we are all praying all the time; we are all taking actions in the world based upon our preferred outcomes, both conscious and unconscious.

Perhaps one of the reasons why it seems that prayer can't fix problems like climate change or terrorism is because, in my definition, we are all praying constantly through our actions. In this light, maybe it is our prayers (through our actions) that are perpetuating the problems in the first place. The majority of our prayers are unconscious actions aligned with the cycles of wasteful consumerism and competitive dominance, so while we may verbalize a prayer for the healing of the Earth, it is through our actions that meaningful history selection is powered and prayer really happens.

As a scientist, I worry that even mentioning the words "faith" or "prayer" will cause some of my readers to throw down the book in exasperation, and I understand that reaction. I am familiar with the historical relationship between science and religion. I myself err on the side of science and discuss faith and prayer very cautiously.

I am a white man brought up in a mix of cultures, including the yogic, Buddhist, and Taoist traditions. These three, the way I learned them, are "worldviews" or even "sciences," in the sense that they do not promote a deity but rather a method for understanding the nature of existence. I have both Jewish and Christian traditions in my family as well, but their influence was primarily of a cultural nature. My favorite authors include the Sunni Muslim Sufi mystic Jalaluddin Rumi, Catholic monk David Steindl-Rast, and physicist Richard Feynman. I have learned about both science and religion and have taken from both fields those truths that align with my inner compass. With this background, I don't bring a religious bias to this discussion; rather, I come with a thirst for understanding.

What concerns me is that the word "faith" seems to have become synonymous with religion, which implies that faith is also limited to religion. But if I don't agree with just one tenet of a religion, am I forced to conclude that my spiritual experiences are a figment of my imagination? To have a religion is to have "a faith," so if I instead choose not to embrace a specific religion this apparently means I also lose access to a sense of faith. That is a lot to lose!

Scientists also have faith; they have faith that, even when their work seems hopelessly confused, the method of science will eventually lead to greater clarity. In this regard a scientific approach is both like and unlike a religious one. Faith in science is explainable by pointing to the laws of cause and effect. A cause may be hidden, yet scientists have faith that there is always a cause to be found. Scientific faith is faith in the process of science. Religious faith is also faith in a process, but that process (as best I understand it, and only in some cases) is attributed to an unknowable god.

Can these two views find common ground? The implication of meaningful history selection is that when we step up to a challenge, helpful events are likely to arise to meet us in that challenge. For example, my wife and I recently needed to find a babysitter so we could go on a date, but all the babysitters and family members we usually relied on were unavailable. With the appropriate intention (anticipated qualitative experience) in my heart and consistent actions to back it up, I felt justified in having faith that the problem would somehow be solved. I expected that the process of meaningful history selection would make a solution more likely. Then Dana remembered having a conversation with a young actress in her theater group whom Ellie had bonded with, and although the woman didn't typically babysit, she was happy to make an exception for Ellie. The moral here, of course, may be more about having faith in my wife than in synchronicity, but luckily I have faith in both!

If my theories are correct, meaningful history selection is a completely rational yet unseen process that spans the gap between religious faith and scientific faith. Religious faith, in some teachings, asks us to surrender ourselves to an unfathomable consciousness higher than our own that may have ultimate control over the events of our lives. Meaningful history selection also affects us through the events in our lives, but in a way that is fathomable. We don't have to throw up our hands and give up a sense of responsibility; instead, we can watch how the world responds to us. Rather than seeking to do the right thing according to an external set of values, we can use the synchronicities of life to learn about the effectiveness of our choices. How does our way of being affect the experiences we receive? Can we find hidden patterns linking how we feel, how we act, and what we experience? To do so requires a profound level of self-honesty. This is my understanding of faith.

I had faith the time I went along with my wife Dana's suggestion to record a music CD for my new baby niece. Together, our purposeful actions catalyzed a fun living-room recording session with twenty family members and resulted in an enduring keepsake.

I had faith the time I trusted Dana's suggestion that we fly to Minneapolis to visit my sister and her new baby, even though there were plenty of reasons to talk ourselves out of it. My parents also ended up visiting my sister the same weekend, and the rental unit Dana found for us just happened to have a pool table and loads of games for kids. The group of us spent an entire day indoors during a snowstorm relishing each other at this special time in my sister's life. How could I have foreseen that? Only through careful listening within the LORRAX process and with a sense of faith in the meaningfulness of life's events did we shape those experiences.

Having faith in my wife's suggestions is a good way to get into flow and find synchronicity. When she made these suggestions, I saw them as singular moments, forks in the road, signals to pay attention. The main hurdle was navigating my own resistance to following those paths. The final decider for whether an event is meaningful to me is me. Still, the beliefs, worries, and other patterns I have built up in my life can make it hard for me to see clearly. By following the LORRAX process, I can continually get better at entering into the cosmic dance with circumstance. I can develop a sense of which circumstances I trust to work out and how I must behave for them to do so.

I know of no better word for this than "faith." Faith is a willingness to step into flow without having all the answers. Faith is being proactively involved in the outcome yet open to circumstances that are bigger than oneself. In this type of faith, one shows up with a plan yet is open to grace.

This worldview implies that every ounce of energy counts. Whether or not we reach our stated goal, we can have faith that the effort put into reaching that goal is moving us forward. If we feel discouraged about continuing in our efforts, faith is a completely practical step we can take. If we reflect inwardly and decide that the goal is something we really want, then knowledge of meaningful history selection can give us the motivation to continue building momentum, even in the shadow of our previous failure. Having faith to keep building momentum long enough to get to an apple is a key part of how our choices can shape our world.

To take this challenge seriously requires a special state of being that is not the norm in the world today. It is more than our intellect can manage alone. We can rediscover a source of power innate in all of us: the power of our heart, which can transform our experience completely.

LIVING FROM THE HEART

As we've seen, from the perspective of meaningful history selection every action has meaning. A purposeful action is one whose meaning is aligned with a coherent intention. Instead of seeking the meaning *of* life, we can seek a meaning*ful* life, or even better, a *purposeful* life that strings all these meaningful experiences into a coherent story. How do we create more meaningful experiences? We take bold actions that reflect our feelings and arrange the apples on the tree so that unique and interesting situations arise. Taking bold actions can be a scary and difficult habit to develop, but living from the heart can help us find the courage to be bold—a courage that many of us don't know we have.

Synchronicity Works for the Greater Good

My friend Tom, a school teacher, once sent me an email describing the process of finding his life path. On the one hand Tom felt he should dedicate himself to selfless causes, focusing on the well-being of others at the expense of his own salary prospects. On the other hand, he was familiar with the difficulty of struggling to make ends meet and wanted to set himself up to be successful financially.

"Neither extreme seems appealing," he wrote. "And then I realized that when I am fully aligned with my highest good, it also benefits others/the whole. There is a sweet spot where my own best interests align with the greatest good for all."

Living purposefully is about finding a sweet spot where both ourselves and our communities are served by the choices we make. If we look around today, we might come to believe that it is an either–or path: be somebody who serves the greater good, or be somebody who serves your own good. Many professions lie near one or the other of these poles.

Synchronicity gives us another way to think about selflessness. It is not based on the global picture (i.e., the career we choose) but rather on our moment-to-moment actions (i.e., day-to-day choices *within* that career). We can use the tree image to think about whether our actions serve just ourselves or those around us as well. If we do something that serves our community, others will find it helpful, and people will be likely to jump in to support what we did. Hence there will be many apples nearby on the tree, and synchronicities will arise to further our progress along the tree in that direction. Our action leads to someone else responding, which leads to more confidence and further action, and the wave builds in strength. On the other hand, if our action is not in alignment with others, then there are not many apples nearby, and fewer people will respond to us. The branch we are on will not expand into lots of branches meaningfully related to our action, and the effects of the action will fade away. Actions that serve our environment will resonate in the tree and take on a life of their own.

Recently I was at a lecture and it became clear that the speaker had no sense of the time. The moderator had left the room, and three hundred of us in the audience were left at the mercy of a speaker who had gone fifteen minutes into lunch. I was involved as an organizer, so I felt somewhat deputized to rein in the situation. Yet I wondered if I would be serving the greater whole. Was this a situation I should try to shape? Would others in the room be glad I had taken charge, or were they soaking up the lecture? Was I the only one stirring uncomfortably in my seat? More and more people kept spontaneously standing up and leaving, so my gut told me I wasn't alone in my perception. I raised my hand for a question, and although he didn't notice me, I spoke up from the audience to say, "Thanks for that point, I think it's very interesting! Also, the session is over, so we need to head out to lunch." I tried to be friendly and assume good will on his part but also to catalyze a change in the room.

If my action was not in alignment with the group, then very few people in the room would agree with my statement, and their attention would quickly shift from me, a minor annoyance, back onto the speaker. Instead, my spark appeared to be on the mark. Polite applause for the speaker erupted immediately—the audience's way of saying "Thanks, bye!"—and most people quickly gathered their things, stood up, and went to lunch. It turns out my action was both selfish and selfless, serving both me and the group. This is a hallmark of being in the flow.

How would this look when mapped onto the tree of possibilities? A branching occurs each time any person in the room decides between staying or leaving. If there were very few people interested in the talk ending, then there would be very few branches on which a lot of people stand up to leave when I interrupt, or very few apples on the nearby branches. (See figure 14.) Since in my case there were so many people just waiting for the chance to get up and leave, there was an explosion of possible branches with apples leading from my action to the end of the talk. (See figure 15.) The spark caught like wildfire and catalyzed a change that others wanted but did not know how to initiate.

FIGURE 14. If I speak up and interrupt but most people want the talk to continue, there are not many branches on which other people respond by standing up and leaving. On most branches, most people stay put and keep listening to the speaker. The apples represent my anticipated experience of having the session end, and in this case there aren't many apples nearby as a result of my action. We would say my action was not in alignment with the intentions of the group.

We could loosely say that actions that align with the greater good are resonant with the tree of possibilities. In physics, a small amount of energy put into a system at just the right resonant frequency has a very large impact. A microwave oven works by irradiating food with microwaves modulated to the resonant frequency of water, so that a small amount of radiation has a big effect on the temperature of the food. In the same way, a small effort headed in a certain direction at a singular moment can resonate with the tree of possibilities and cause a big change.

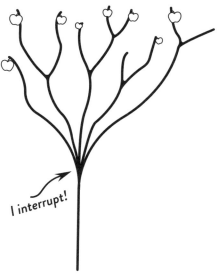

FIGURE 15. If my interruption aligns with the way many other people feel, then there are many nearby branches in which many people in the audience get up and leave. The apples here represent my goal of having the talk end, and in this case there are quite a few ways that can happen. We would say my action was in alignment with the intentions of the group.

This doesn't mean that serving the will of the group is inherently a good thing. Mob mentality can be very destructive, and resonating with the tree can have negative effects. Even in a mob scene, there are singular moments at which a person who goes against the grain might be able to awaken the better nature of the individuals in the mob and be a spark that brings sanity to the situation. We can light sparks that resonate constructively or destructively. The process of meaningful history selection doesn't discriminate between these two. To know whether we are in alignment with the greater good, we have to develop our inner sense, the alignment between our head and our heart. This alignment is our guide toward finding actions that serve the community for the better. Harmony is achieved when these two influences support each other.

Now, I don't suggest we should always look to circumstances to validate our choices. That's a slippery slope, and we may find ourselves always trying to get positive feedback from the world rather than being aligned with our purpose. Rather, we can note whether there was some positive response to our action. Sometimes the response is direct, like a woman who said "Thank you!" to me after I broke up the meeting. Sometimes the response is indirect, like the simple fact that the audience responded positively. Or the feedback could just be a feeling of greater empowerment or clarity inside ourselves. The change doesn't have to be big. If our action comes from our heart, it will increase love in the world.

Returning to my conversation with Tom, I gave him a hypothetical example of passing Madonna on the street. If I am too shy to introduce myself, I get no more than a passing glance from her or others in the street. But if I break out in my own a cappella version of "Like a Virgin," she may be entertained, and people in the street will also find themselves engaged by it. This bold action may resonate with people's yearning to see something unique and exciting, and many branches emerge in which interesting events happen. Maybe everyone claps and cheers, maybe Madonna asks for another song, maybe somebody in the street invites me to perform at their cafe. These events arise from the spark I lit by being bold.

Tom wrote me back to say that at first he'd thought my Madonna example was kind of ridiculous. The day after he received my email, he and his wife went on their first outdoor walk with their infant daughter. They deviated from their normal routine and ended up at a cafe for lunch. After sitting down

Tom realized that the musician Carlos Santana was dining a few tables away. The moment felt auspicious, and our conversation came back into his mind. He thought about asking Santana to meet his baby daughter, but he ultimately convinced himself that leaving the rock star in peace was better for everyone.

But was it? As soon as the opportunity passed, he began to regret not seizing the moment. He had seen the opportunity, but hadn't been able to venture outside his comfort zone to light a spark. If he had taken the chance to say hello, might branches have emerged that served others as well? Maybe Santana also had a new baby granddaughter who lived far away, and he would smile to be reminded of her. Or maybe somebody else at the table had a friend with the same name as the baby, and a meaningful conversation would be catalyzed at their table. Certainly it would be a meaningful story to tell his daughter when she was older. If Tom had followed his heart, meaningful history selection would likely have led to circumstances that went beyond what he had consciously intended. Synchronicity connects us and benefits everyone in unexpected ways. It is nature's way of reminding us that we all need each other.

But we are not all able to be bold all the time. Are we then doomed to mediocrity and isolation? It was despair over this question that brought me to discover a new way of seeing the world: through the lens of my heart.

A New Pair of Glasses

The process of lighting a spark or stepping out of our comfort zone can feel threatening. Gathering the courage to speak to a celebrity in public is difficult. Sometimes it's so overwhelming that it is futile to try to overcome it. For me, being bold is like a one-two punch. First I feel afraid that if I go outside of my comfort zone—in this case, talk to the celebrity—I'll be humiliated. Then I feel ashamed that if I *don't* do it, I'll never have the courage I need to reach my potential. I'm trapped! How likely is it that this miserable process will ever lead me to a useful action? That particular dilemma may not apply to everyone, but the general experience of being trapped between bad options may feel more familiar. It comes down to the framework we are using to judge life's events. I would call this the ego's framework, and it shapes many of the events of our lives by default.

The ego is like a pair of glasses that interprets every event through the same filter, coloring our life with comparison, analysis, and separation. When we see the world through this pair of glasses, life might seem filled with opposing bad options, like when Tom was deciding whether to introduce his daughter to Santana. His choices apparently were either to be embarrassed or to be ashamed. But the glasses of the ego do not show us the one, true reality; they display a version of reality that reflects their own filter of separateness. Tom's imagination of how Santana would respond was probably not accurate. When we see the world through this lens, we create a reality of separateness. If Tom acts out of separateness, meaningful history selection brings experiences that keep him feeling separate. He might be thinking of saying hello, but he hesitates a little too long. Suddenly somebody else walks up to talk to Santana, which reinforces Tom's narrative that he shouldn't go pester the celebrity any further. The ego is constantly looking for evidence to support its framework, but since it sees the world through that framework, everything that happens can be seen as evidence of separateness.

There is a different pair of glasses we can wear: the lens of the heart. The heart—in French, *le coeur*—is a source of "courage" (note the etymological similarity) that naturally dissolves

A Song with a Message

"I was doing errands and a song came on the radio that I hadn't heard in a while. In the middle of the song I parked the car and went into a store. I remarked to the cashier that they must be listening to the same radio station as me, because the same song was playing and was at the same part of the song. The cashier told me they were listening to prerecorded music, so it was a coincidence. This caught my attention. I paused and listened more carefully as the chorus came on. Upon reflection, the lyrics contained some insights for me that I think I needed to hear that day!" (Story contributed by Julia Mossbridge)

the fear promoted by our thoughts. The tactic of the heart is to shift focus away from what we fear and onto what we love. For instance, the feeling of caring for his baby daughter could dissolve Tom's fear of embarrassment. Running through the situation in his head only got him more entrenched between the only options he could see—embarrassment or shame. By shifting his focus onto his love for his infant daughter and the joy of sharing this experience with her, he might have reclaimed his creative energy and spontaneously known how to proceed. The entire conversation around comfort zones becomes irrelevant when the heart focuses on what it loves.

I learned to see through this new lens when I was facing my own internal conflict. Although I had spent decades trying to overcome my fears of breaking out of my comfort zone, it had not become any easier. I thought for sure that with enough practice I would get to a place where speaking up in my family or performing in public came naturally. That was not the case.

Every opportunity brought up the same anxiety. It was as if I had learned nothing from my previous experiences.

In this particular opportunity, I had been invited to the Burning Man festival, but I declined because it was my daughter's first week at a new school and I wanted to be home to support her. Thankfully, it turned out to be an easy transition for her, and I realized on Saturday that I could go to the festival for the final day.

Earlier that morning, I had stepped on my glasses in the dark, and they had snapped cleanly in two. I ordered a new pair online but found myself temporarily without a good pair of glasses. My "filter" was apparently being rebuilt.

Dana and I talked through the Burning Man decision. It was not an easy conversation, for there would be a significant impact on my family. What's more, it was a six-hour drive, and I didn't have a ticket. I wanted to trust in the responsive cosmos to solve that problem for me, but the risk of arriving in the desert and not being allowed into the festival felt so daunting. I struggled to overcome my fear of taking that risk. A secondary layer of fear told me, *If you can't trust the cosmos to solve this problem then you aren't living your message.* Oh no! My ego framework was overwhelmed. I tried to just sit with the discomfort of this impossible choice without having to do something about it.

In this open-hearted state, I remembered that I had planned to attend a new church community that morning, a spiritual center I wanted to become more familiar with. Since I was not yet driving to Burning Man, I thought I might as well head across town and join the congregation. I did, and what I experienced there changed my perspective permanently.

As a white person, I am beginning to wake up to the role of white privilege in my life. Part of this involves recognizing patterns of behavior that I have inherited from my white culture—patterns that may be very visible to others from the outside but that I often can't see. The church congregation I was headed to is a majority black community. In the sermon that morning the minister shared about seeing the world through the filter of the heart. I connected some dots and recognized that the whole community seemed to live from the heart very authentically. It was a trait I had seen before, but now I saw it in a new light.

I observed the speaker who, despite some apparent physical limitations, shared his message with calm confidence and determination. If I had felt as exposed and raw as the speaker was, I would likely have felt self-conscious. Yet I noticed that the congregation witnessed him exactly as he was, without judgment or impatience. I also noticed that one of the assistants didn't hesitate to yell out from the audience or sing a melody here and there. He wasn't playing a specific role in the service; he was just sort of being a part of things. I observed that many of the congregants were quick to stand up and dance and sing, expressing themselves with natural ease and at the appropriate moment, in contrast to the people I grew up with, who would have just been observers, not participants.

I realized that these people were seeing the world through their hearts. While I often take the risk to stand up and dance or sing authentically, I had always come from my head, so I was usually left wondering, *Is this the appropriate time or place?* I saw now that when the action comes from the heart, that question becomes meaningless because the suffering of separation and striving for perfection drops away.

I have sometimes made unwise judgments about people who were heart-centered, seeing them as careless or not serious. I was once the bandleader in a group where I was one of two keyboardists. Most of the ten people in the band were black, including the other keyboardist, a man named Paul. I judged Paul to be kind of flaky, and I felt like my keyboard playing was better than his. But then the manager of the band advised me to learn from Paul, because he had something I needed more of: feeling.

My judgments came from a limited mental perspective that betrayed my own sense of separation and inadequacy. My insecurities were likely obvious to those seeing through the heart-centered lens. All the other people in the band probably saw it clearly and felt sorry for my angst, but I couldn't see it. In striving to compare myself to Paul and lift myself up, I was left feeling disconnected from my community. My biggest regret looking back on that experience is the missed opportunity to be a part of that community, to feel connected and respected and loved.

Yet on this day in this particular spiritual community I saw the heart in action and was ready to make that shift. It was the only way out of my mental dilemma, which had built to paralyzing proportions with my decision around Burning Man. On this morning I had found a new way of seeing the world in which the courage to act came naturally and my struggle dropped away. It was an invitation to a way of shaping the world effortlessly (although unpredictably) with love, rather than pushing my agenda exhaustively. I sang, danced, and socialized, unperturbed by how I might be perceived. My heart focused on the joy, and there was no more hesitation.

From my new perspective, I recognized this heart-centered trait in many of the black people I had worked with or been friends with over the years. I think I have only scratched the surface of understanding the differences between white culture and black culture, and I may have it all wrong. For me, the heart-centered way of being came from reaching a place where my mental attempts to try to "make it" in the world had become futile. No matter how hard I pushed, I wasn't going to get where

I wanted to go. I was up against a brick wall. This "opening to grief" awakened in me the courage to be myself, regardless of what others may expect of me. There was no use in maintaining my facade, for it was not effective. I finally felt the power of Zukav's words: "The road to your soul is through your heart."[90]

The Irony of White Privilege

I had also been reading *Waking Up White* by white author Debby Irving, a book intended to help white people understand white privilege. It occurred to me that maybe systemic oppression—something the black community experiences every day—can have this same effect of causing one to drop the facade. For many black people in American society, it doesn't much matter how they act because white people prejudge them. Black people are forced to accept that they will never be accurately understood by white people, and the economic system we live in is structured to treat them with suspicion regardless of their personal qualities. This brick wall certainly seems to me like it could make a person say, *I'm just going to be myself as best I can within this oppressive environment.*

I have never been the target of systemic oppression. My white privilege affords me the belief that if I just keep trying harder, keep working to impress the right people and do what people say they want of me, then I can move up in the world. I can have a bigger career, make more money, become famous. It is precisely this attitude that has kept me stuck in my head, trying to outdo and overcome the external standards the world has set. I learned from a friend that for many black Americans, church is the one place they can drop the facade and release the outrage and humiliation caused by living another week in white America. The depth of frustration gives rise to a commensurate depth of expression.

As a white person, I rarely get pushed down so hard that I am compelled to find out who I really am. I am not usually forced to dig deep inside myself to survive. My white privilege affords me the belief that I can reach the standards of white society. Instead of opening my heart and connecting, I isolate myself and try to conquer the world. I bang people over the head with my message so they will get how important it is. But with my new lens, I see that speaking, acting, and singing through my heart provides me a different way to truly connect with others, making my work more meaningful for me and for them.

Absurdly, our society (whose official structure is created largely by white people) had trapped me (a white person) in a state of mind that was holding me back. Paul had something I couldn't even recognize, yet we can all feel. It is clear to me that my ability to reach my true potential has been limited by the very systemic oppression that I benefit from—a tragic irony. The truth is that everyone suffers when one group of people is oppressed. Conversely, we all benefit from an understanding of the experiences of people who are different, not just in a holier-than-thou sympathetic way, but in becoming whole ourselves, or "whole-ier *with* thou." White people experience pain from racial inequity as well, just as men experience pain from gender inequity, although they may or may not notice it. It is a pain that comes from separation. There exists an undercurrent of insecurity within white culture, and similarly within masculine culture; we hold the reins of control, so we must maintain the facade of confidence even though the world is fundamentally uncertain. To the extent that we trap ourselves in having to look like we know better, we live a lie and are unable to experience authentic connection. The fight for equity, whether racial or otherwise, is not only for lifting up the oppressed; it is also important for liberating the spirit of the oppressors.

I am grateful for the teaching that the individuals at my new spiritual community have provided me on my own journey toward authentic expression. Later that week, my new pair of glasses came in the mail.

How does this relate to synchronicity? This new pair of glasses allows us to align our head with our heart. Our heart's feelings have the effect of reaching into the possibilities and selecting futures that align. For instance, I might feel that I want to learn everything I can from being in my band. This populates a bunch of apples on the tree, including one branch in which Paul is hired as a second keyboardist. But if the thoughts in my head don't align with this feeling, then I am likely to choose actions that sabotage the experience. I end up being in a tug-of-war with Paul, and I don't achieve the deeper experience of connection I want.

When we allow our head to see what our heart sees, our ego's choices align with our heart's intention, and meaningful history selection is likely to bring us useful circumstances. Learning to see with my heart was a breakthrough on my path toward healing my fear of "being seen as I am" and has made me more able to step into the unknown to experience synchronicity.

Selfless Synchronicity

While being seen as I am has been an important healing for me, it has paradoxically led me to focus more on others. This is a virtuous cycle, because when I direct attention onto others I tend to invite more synchronicity, which benefits everyone.

I have heard some self-help teachers suggest we should "feel the fear and do it anyway." With my old pair of glasses, the clear message is, *Conquer our fear.* What a daunting command! Yet through the lens of the heart, there is a gentler way. Sometimes we can find courage by attuning to others in their struggle. Do we notice someone else holding themselves back, afraid to speak up or go for it? We can use our courage to bring about meaningful experiences for them. We receive a gift from focusing on the well-being of others: our fear dissolves. When we act on behalf of another person, we are no longer worried about ourselves. It's like the other person gives us an excuse to stand out, or at least someone to stand *with*.

Recently my mother was visiting town, and we set up a dinner date at a pizza place called the Cheeseboard. At the last minute she brought along my nephew, Kiva, a very talented musician. Kiva has been performing in public since he was five years old, but now at fourteen he is more self-conscious. As I walked into the restaurant, I noticed a small band playing, and per my usual practice I took a moment to introduce myself to the band. The bandleader invited me to play a few songs during intermission. I was happy for the invitation but also nervous. I sat down with my family to enjoy some pizza first, and in talking with Kiva and my daughter, Ellie, I learned they were both open to singing with me. Suddenly I was more at ease—now I had a band! At least, I had people to stand up there with me and be in the spotlight. This made my own apprehension more tolerable.

When the intermission arrived, the kids were a little nervous about choosing a song. I remembered that they had sung "True Colors" by Cyndi Lauper together before, so I suggested it. Kiva and Ellie took the spotlight as they traded verses and sang together on the chorus, "I see your true colors, and that's why I love you." It was a wonderful moment for everyone in the room, because who doesn't like to see cute kids performing well in a restaurant? It was a symbiotic exchange: they gave me a sense of purpose, and I carved out the spotlight for them.

Afterward, we sat down and finished our pizza. The experience was gratifying for me as an aspiring musician. But it also turned out to be an example of how living in the flow benefits everyone. The next day I saw my mom again, and she gave me the backstory. My brother had been trying to get Kiva to play at the Cheeseboard for a long time. They were delighted he had finally done it! In my mom's words, "It wasn't going to happen until Uncle Sky paved the way."

I am often concerned about taking up too much space, perhaps because I grew up between two different homes with five siblings and five parents combined. It hasn't been easy to learn to step into social situations boldly. But in this case, I ended up acting in service to something bigger than myself—a transformative experience for my nephew. How was I to know that in the moment? Often, I can't. I just trust that when I follow my heart and live my life authentically I am also serving the greater good.

This is what I mean by relying on inner knowing to tell where the meaningful opportunities are. When my friend Tom says there is "a sweet spot where my own best interests align with the greatest good for all," he is describing something important about synchronicity: it seems to bring good into our own lives as well as the lives of those around us. If we want to make the world a better place and bring more fulfillment into our own lives as well, synchronicity and flow are great ways to do it.

For an easy way to practice this, notice the dynamics that occur the next time you have to choose a restaurant with a group of people. This can bring up a mix of input from people who may have dietary restrictions or different preferences regarding cuisine, location, ambience, or cost. Yet this decision is not usually a critically important one, so it gives us a chance to experiment with the balance of assertiveness and receptivity that shows up in the LORRAX process of listening, opening, reflecting, releasing, and acting. Notice any particular options that arise that may not at first seem good but might lead to a great experience nonetheless. Notice how you might listen and reflect better, or perhaps act more definitively, to get the group into flow. We can ask ourselves, *What is needed in this situation? How can I create more joy, love, laughter, or constructive energy in this situation?* Others benefit, and we do too.

When we start to pay attention to synchronicity and flow in our lives, whether through the LORRAX process or some other method, we may find that they appear all over the place. Experiences that arise in this way are destined to help us grow, but not necessarily to bring us what we want. The more we accept opportunities to engage with flow and synchronicity, the more we will notice that we sometimes miss those opportunities. This brings about a paradoxical relationship between abundance and missed chances.

An Abundance of Mixed Blessings

I want to catch every opportunity, but I don't. I find it difficult to stay as present as possible, to be as bold as possible, ready at any moment to do what might be necessary to bring about meaningful situations. This awareness can sometimes cause me angst. Through practice, I now find that I can gracefully let go of opportunities I miss. Recovering from one missed opportunity is necessary to be ready for the next one. To let go and move on, I acknowledge the feelings of grief over what I have missed. I am willing to tell myself the truth about what fears may have held me back. I accept that the thing I am attached to might not come back again.

Recovering from missed opportunities is like restorative justice for our soul. Restorative justice seeks to reconcile victims and perpetrators by following four steps: inclusion of all parties, encountering the other side, making amends for harm, and reintegration of the parties into their communities.⁹¹ Living authentically is a path of soul justice. When we miss an opportunity to align with our purpose, we have done harm to our soul. Now, I do understand (and I hope you do too) that satisfying the soul's purpose all the time is probably not feasible. We will always be making "mistakes" in this regard, so we can be humble about it and give ourselves a break. Still, the steps of restorative justice can be useful to help us reconcile with our soul's purpose when we miss an opportunity.

When I was leading a band and allowed my judgments to get in the way of learning from Paul and connecting with the other band members, this was a missed opportunity to align with my soul's purpose. The first step in restoring justice within myself is to speak truthfully to myself about the missed opportunity. This includes my soul in the conversation as if it were a stakeholder in the situation. It took me many months or even years to

---

Being Someone Else's Serendipity

"In celebration of my last days in Bangkok, I hit up the nightlife, which was unusual for me. At about 3 a.m., I left a club and went off the beaten path down an empty street to grab a cheaper taxi home. I was quite far from the club when I heard the telltale sounds of uneven, clumsy walking behind me. I was a little worried about my safety, yet when I turned around I realized that I recognized the person! It was a friend of mine, stumbling home after a day gone horribly wrong. He needed a friend, a good listener, and amazingly, the universe gave him one in me." (Story contributed by David)

acknowledge to myself that my attitude was a part of the problem I had with Paul.

When I think sincerely about why I missed the opportunity, I encounter my soul face to face without defensiveness. Honesty with myself allows me to grow from the experience so I am prepared to act differently when a similar situation arises in the future. In dealing with Paul and the band, it was difficult to realize my own limitations in relating to people who are different from me. In finally acknowledging that an opportunity had been lost, I was able to feel authentic grief over the experience. This allowed me to make amends by recommitting to learn what is necessary about myself—in this case, how to respect others and myself at the same time—so I can do better in the future. Finally, I reintegrated into the community by lifting my head up and looking for the next opportunity.

To be clear, I am talking about restorative justice with *myself*, not with Paul or the band. Whether I choose to heal the mistakes of the past in the world outside is a separate issue. But the process of internal restorative justice can help us get back into flow in our own lives.

As whole people, we are bigger than any one desired opportunity. When we focus on making amends with our purpose or our soul, we see that more opportunities are coming. The responsive cosmos is a cosmos of abundance. There is no end to the universe's process of bringing us meaningful circumstances; we couldn't stop it if we tried. The shadow side of this fact is that, in a universe where unique snowflakes of opportunity are falling from the sky, there is no way to catch all of them. We catch whichever blessings we can hold as they fall onto the ground like a carpet of flowers. They are not ours to own; their profusion teaches us about impermanence. Every circumstance involves aspects of gain and aspects of loss, aspects of joy and grief. What if life is not about scrounging for a crust of bread or a 401(k) but about recognizing and managing an abundance of mixed blessings? Life can only give us more opportunity than we can handle if we can refrain from beating ourselves up when we can't catch all of it.

Some people feel that living synchronistically is just a Pollyanna attitude. The world is full of dubious situations, so looking to the world for useful guidance seems like wishful thinking. But a *responsive* cosmos is not a *friendly* cosmos. It is a mirror. The goal is to find inner purpose, not to get what we want. When we live with purpose, we might earn money and power, but these are only useful to the extent that they serve our purpose. Otherwise, they are distractions. To live synchronistically is to welcome the seemingly positive and seemingly negative experiences with equal suspicion. Everything is a mirror; therefore we are interested in the experience only insofar as it teaches us about ourselves.

Living synchronistically doesn't mean that difficult experiences don't arise. Over the past few years I have had a series of difficult incidents that seem to repeat themselves with the homes I have lived in, one of which was relayed in chapter 2. Handling these occurrences has forced me to develop skills that I find very difficult to acquire and master. These troublesome experiences are just as synchronistic as the more positive examples.

Synchronicities reflect the condition of our soul. They echo both the authentic emotional experiences of our heart and the intellectual choices we make. I think of synchronicity not as a walk in the park but as the navigation of a mountain pass. We can find great joy even in the danger of a mountain pass, but if we persuade ourselves that the responsive cosmos is a walk in the park, we might find ourselves caught off guard by unexpected trouble. A synchronistic mind-set is therefore on the lookout for mixed blessings. We catch the opportunities we can and gracefully let go of the rest. What a relief! Sometimes I hold onto life as a thing to be captured and held onto, showcased to others (or even to God) when I die: "See! Look what I did!" But as a catalyst of experiences rather than an owner of them, I can only catch a few of the snowflakes and truly appreciate their beauty. I feel a bittersweet joy for both the wonder of abundance and the grief of loss.

Wonder and awe are natural reactions to this mixed abundance and a great way to get back into a mind-set of flow when we realize we have missed an opportunity. Psychologists have reported that awe has many beneficial effects, including increased generosity, reduced sense of entitlement, ethical decision-making, helping behavior, reduced impatience, and increased humility.[92] Researchers Christina Armenta and colleagues say that awe creates an "upward spiral between happiness and positive life outcomes."[93] It certainly stands to reason that meaningful history selection, if it is shown to be an accurate view, would enhance this upward spiral by amplifying the likelihood of meaningful experiences.

In *The Bell Jar*, Sylvia Plath describes the constrained abundance afforded by the tree of possibilities:

> *I saw my life branching out before me like the green fig tree in the story. From the tip of every branch, like a fat purple fig, a wonderful future beckoned and winked. One fig was a husband and a happy home and children, and another fig was a famous poet and another fig was a brilliant professor, and another fig was Ee Gee, the amazing editor, and another fig was Europe and Africa and South America, and another fig was Constantin and Socrates and Attila and a pack of other lovers with queer names and offbeat professions, and another fig was an Olympic lady crew champion, and beyond and above these figs were*

many more figs I couldn't quite make out. I saw myself sitting in the crotch of this fig tree, starving to death, just because I couldn't make up my mind which of the figs I would choose. I wanted each and every one of them, but choosing one meant losing all the rest, and, as I sat there, unable to decide, the figs began to wrinkle and go black, and, one by one, they plopped to the ground at my feet.[94]

To thrive in an abundant cosmos requires us to feel the grief of our losses, because an opportunity to go in one direction often precludes another direction. Tom's earlier dilemma, choosing between careers, was a valid one: he couldn't have every fig. The beauty—and mixed blessing—of living in an abundant cosmos is that there is an equal profusion of gain and loss. It seems that our choice is between a comfortable but mediocre life without losing or winning, or an abundant life filled with both joy and grief. In my experience, seeking only comfort leaves an empty hole inside. I think that deep down we long to live vibrantly. We want to experience whatever bounty the responsive cosmos throws at us.

But wait! Is the cosmos really abundant? It certainly doesn't *feel* abundant sometimes … but why not? Why does it sometimes feel like we have to fight for everything good? Why are we limited within the tight constraints of bringing in enough money and having enough time? These constraints are real. Abundance is not about having fruit dropped in our lap; it is about being in an orchard with fruit around that we can pick. We don't all have the same fruit available to us, but we all have opportunities to improve our present circumstances. The accessible fruit for me on one day may just be finding the strength to get out of bed in the morning, whereas on another day it might be speaking to a congregation of world leaders.

With an honest and accurate picture of our present constraints, we can plant seeds and light sparks that set synchronicity in motion wherever we are. But we can't harvest the whole orchard. We may not even be able to do more than pick one or two pieces of fruit. Is that enough?

I think so. Seeing all events as mixed blessings helps us see the meaningfulness even in difficult situations. Rather than trying to avoid problems, we can immerse ourselves in a synchronistic ocean of purpose. The responsiveness of the cosmos is a mutual relationship. It provides part of the potential for fulfillment and growth. Yet first we must plant the seeds and set the direction. And after those potentials appear, we must stand up, find a stool, and start picking the fruit of mixed blessings. The responsive cosmos always fills in the second step, but the first and last steps fall to us.

How do we find the courage and strength to find that stool and keep reaching for fruit? Disappointment with what the responsive cosmos brings us can be a big obstacle to living in flow. Is this piece of fruit enough? Other people often seem to have earned more or accomplished more than we have. The world, and life, just don't live up to the advertising. Csikszentmihalyi writes, "Whenever some of our needs are temporarily met, we immediately start wishing for more. This chronic dissatisfaction … stands in the way of contentment."[95] So how do we get over the hump of expectations in order to see the mixed blessings in our lives as synchronicities?

Gratitude

Gratitude is a powerful tool for creating more synchronicity and getting into flow. I find myself more likely to accept the opportunities that show up in my life if I feel grateful, because gratitude reduces my expectations and opens my eyes. In a state of gratitude, I am more likely to see the flow that is right in front of me.

One reason why it can sometimes be hard to feel gratitude is that it involves feeling grief. In the last section we talked about grief as a necessary part of staying in the flow, and this relates to gratitude. In my experience, it is hard to be grateful when I can't allow myself to feel sad for what I have lost. Gratitude grows out of grief. Gratitude is a feeling that emerges when I see what I've lost and also what I still have. For me, gratitude comes from a recognition that there is always room to fall, always more to lose. Preserving what really matters in my life requires a realignment of priorities.

> ### Trusting That Urge
>
> "My dad was diagnosed with lung cancer while I was living across the country from him. I hated being so far away and planned for some time off work to visit him. As the date was approaching, although my dad's treatments had become routine, I had a strange feeling that I should go back sooner, so I moved my flight up a week. The day after I arrived to see him, he was admitted to the hospital with sepsis. One day later, he died. If I hadn't followed that feeling, I would not have been with him when he passed from this earth." (Story contributed by Michelle)

Research is being done on how negative emotions such as grief may serve as powerful motivating factors. Based on their research into the positive aspects of gratitude, Armenta and colleagues recommend that future studies explore "the ways that negative feelings that accompany gratitude may uniquely boost motivation to engage in positive action. For example, after reflecting on how much a parent or mentor has supported them, lingering feelings of guilt or indebtedness—when not chronic or overwhelming—may light a fire of change and impel individuals to work harder or be better people."[96] The pain of a lost opportunity can bring us gratitude for what we have and motivate us to act with more courage the next time an opportunity arises.

In 2010 I applied to be a speaker at the Science of Consciousness Conference in Tucson, Arizona. I really wanted to be offered a speaking slot so I could discuss my work with a wider audience. My primary motive was to share my work so it (and I) could be appreciated, but my deeper motivation was my feeling of separateness. I felt I had to work to convince others to see things my way; I needed them to step out of their own skins and into mine.

Instead, the conference organizer jumped on the fact that I am a musician and asked me to coordinate an evening of entertainment at the conference. This wasn't what I had asked for. I wanted to be seen for my research but was being offered a chance to be seen for my art. I planted a seed only to find some other strange plant growing out of the ground! In a responsive cosmos, the seeds I plant are not literal demands—they are metaphorical demonstrations of my soul's love. The cosmos responds not to the specific experience I think I want, but to the symbolic experience my heart is seeking.

I accepted her challenge and have become a regular member of the volunteer team every year since. The experience has brought me many valued friends, new connections and colleagues, and even a job through one of those connections.

Gratitude is not my default reaction to life. Often I approach situations from the perspective of what I need out of them. I unconsciously treat others as the supporting actors in a play titled "Me, Getting What I Want." Sometimes I feel such a gulf to fill that I think I'm entitled to certain experiences. What do I miss when I feel entitled? My desire to be a speaker almost made me overlook the amazing opportunity I was offered. I had to look deep inside myself and ask whether I was comparing myself to all those other people in the program so I could feel good about my accomplishments.

If I am honest with myself, my desire to speak at the conference is a surface-level priority. I can peel it back to see more important things waiting for me underneath. Gratitude allowed me to realize that the conference organizer appreciated something special about me. By being grateful for that, I was then able to open my eyes to what the moment was offering. Beneath my desire to be appreciated lay my desire to be included. Being included as an integral part of the organizing team gave me the gift of community and belonging. I didn't get what I thought I wanted, but upon reflection I realized I wanted what I got.

Not everything is possible. Synchronicity doesn't typically solve our problems the way we want them to be solved. To experience more synchronicity, practice seeing what is being offered. The responsive cosmos tends to bring us experiences that reflect our deepest values, but when we are only aware of our top-level needs, circumstances can seem out of alignment. This can set up a negative feedback cycle. For instance, out of frustration at not having a chance to speak, I might act in ways that alienate me from the organizers, which would sabotage potential future synchronicities. Even though I am disappointed now, there are many available future branches that could be very satisfying. If I am not graceful in handling the situation, I may trim those future possibilities off. (See figure 16.) By contrast, an attitude of gratitude opens my eyes and my heart to the abundance of mixed blessings that may be close by on the tree but haven't appeared quite yet. (See figure 17.)

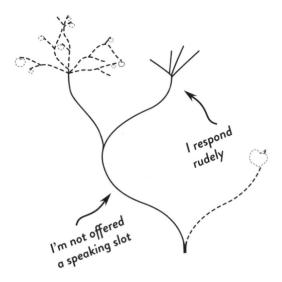

FIGURE 16. If, in my disappointment, I respond rudely to the organizers because I missed the apple on the right, I make it practically impossible for meaningful history selection to create a future opportunity with these people. I trim off all the branches (on the far left) that have apples, representing future circumstances that might have arisen in which they would have called me. On all the remaining branches (in the middle), even if a good opportunity were to arise, the negative interaction we had will color their opinion, and they won't decide to call me. These branches do not have any apples.

Gratitude is about uncovering what we truly want at our core. We may have to dig deep to recognize what is superfluous and then prioritize what really matters. Gratitude allows us to ignore the little things that we might habitually gripe about in order to reach for those truly soul-satisfying experiences. The washing machine breaks. We don't have the ingredients we need for dinner. Our appointment is canceled. Our child spills a drink on the couch. So what? If we analyze the actual consequences of this event and compare them to what gratitude tells us about what

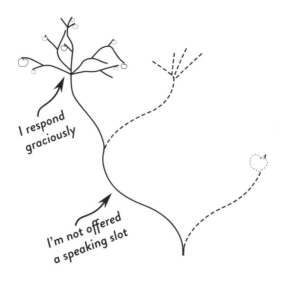

FIGURE 17. If I respond graciously to a disappointing situation (the left branch), I land on a branch with many future apples, even though I missed the apple on the right-hand branch.

really matters, we can release control of things that are of no real consequence. Our attention can flow into activities that move our life forward in the most important ways.

If you are annoyed by little things, then maybe you aren't focused enough on your purpose. Are there any things in your life that you complain about that are not of central importance for your life path? If so, what does it feel like to shift your attention back to your most authentic purpose? Do you forget the minor annoyances? Do you reclaim your energy?

6

AUTHENTICITY AS FLOW

For most of my adulthood I understood gratitude as an attitude I had to conjure. I had to override feelings of disappointment and say ridiculous things to myself, like "I'm grateful that I didn't hear back from the event organizers about booking that gig because it will force me to improve my marketing materials," or "Thank heavens I missed the bus this morning, because it will teach me to be more patient with life." Gratitude often didn't feel authentic, so it didn't become an integrated part of my daily life. It was something my head was saying to override a more authentic feeling: chagrin.

Can I feel more gratitude if I just force myself to be a better, more self-aware person? Although I see that greater self-knowledge improves the quality of my life, that doesn't mean I'm willing to sit on a mountaintop or meditate in a cave for years on end to become self-aware. If that's what it takes, forget it, I'll do the best I can with what I have. I'll have to figure out my path to enlightenment within the context I have available, which for me includes career, marriage, parenting, and enjoying my life.

Living in flow is less about what I do than about who I am. Awakening to who I am involves getting into the natural rhythm of what it is like to be me. This flow requires something so simple yet so difficult that it is more like a life path than a step-by-step checklist: I need to rediscover my authentic self. I think each of us can awaken to your own natural rhythm from within your lives, without dropping out of them. When we find our natural rhythm of life, gratitude and the other benefits of flow may come more easily.

Walking an authentic path means basing my choices on what really works for me, as opposed to what others think I should do. Basing my choices on what I really want and feel isn't permission to be selfish or narcissistic. If Dana wants to go camping on vacation and I want to go to Disneyland, out of my love for her I may tune into a deeper desire for her to get what she wants and enjoy herself. I would choose camping because I authentically want to bring more joy into my family, even though it is not my number one personal preference. Being authentic doesn't necessarily mean getting my way, but it does mean knowing what I want, being honest about it, and choosing consciously.

If you are in a position of influence, whether in business, politics, education, or otherwise, here's my message: stay there! You are needed in that role, not sitting cross-legged on a mountaintop. When we talk about self-knowledge, that just means being your authentic self. That can be a challenge if your authentic self is buried under many layers of rubble. The rubble can take the form of automatic reactions and learned emotional responses that cause you to see situations through a warped lens. As you remove the rubble, you may find your new pair of glasses.

Finding Your Authentic Self

Your authentic self is who you would be if there were no other factors persuading you to be different. Think of it this way: your authentic self emerges as you discover your habitual fears and un-choose them. Let's call these habitual fears "ego habits." For instance, in meetings at work or school, do you ever second-guess yourself and refrain from raising your hand, even when you have something genuine to say? If so, I suggest this is an ego habit you've adopted over time, maybe because it helps you feel safe.[97]

My ego habits usually arise out of a certain type of fear. Maybe it's fear of saying the wrong thing and then being fired. Maybe it's fear of failing a class, being laughed at, or being found out as a fraud. These types of fears hold many of us back from being ourselves. Not all fears are this way; some are in fact quite useful. I like to distinguish between imaginary fears that hold us back and useful fears that push us forward. A fear that pushes me forward is the fear of missing my publishing deadline. It is a fear about something that has a realistic chance of happening![98] It's a useful fear for me because it gives me energy to overcome my resistance and sit down to work. This type of fear eventually leads to a positive result: getting my book completed. In contrast, ego habits are imaginary fears about things that will probably never occur, and they often cause us to behave in ways that hide our authentic selves. Being afraid to raise your hand in a meeting at work may stem from a childhood experience that really did happen. However, most adults who raise their hand to speak in a professional setting would not be laughed at or fired.

You can experiment with the following practice. Note a time when you feel a fear that stops you from doing something. Then, if you have the courage to do the thing anyway, look back and determine whether your fear came true or not. If your fear did not come true, it's an opportunity to reprogram yourself. Close your eyes and imagine being right back in the moment of fear and worry. Let the feeling back into your bones; allow yourself to feel it. Then remind yourself that it didn't happen. By mixing the visceral emotion of fear with the new knowledge that no problem actually occurred, you have a chance to retrain your body to be less reactive to that emotion the next time it occurs. The stress hormones flooding your system in response to your fear are less likely to take over because your body not only knows but *feels* that they aren't called for.

I encourage you to do this a few times and write down your results in a journal. Keep practicing, and track your results. Over time you can build a new body of evidence that convinces you that your fearful thoughts are not always accurate at predicting the future. Zukav offers a warning in this regard: "When you choose to challenge and to release a negative aspect of yourself, that aspect comes to the foreground."[99] In other words, meaningful history selection will likely bring you lots of opportunities to practice your new way of being!

Becoming authentic requires controlling our thoughts so we don't automatically obey our ego habits. Our authentic self knows what to do. It emerges when we aren't compelled to follow where our ego habits lead us. Having more control of your own mind in this way is what we mean by mindfulness. Who wants to be at the whim of a reactive program inside their own head? Who wants to abandon control of their own choices? Being mindful doesn't mean we never listen to the voice of caution or doubt; but we no longer feel as if we have to do what it tells us. We are aware of having a choice. The result of self-awareness and mindfulness is to perceive our motivations clearly so we can make choices that actually benefit us in the long run. When we do this, our authentic self reemerges.

Even as we become more aware of our authentic desires, the ego never goes away. But that's fine; we need the ego's advice. Its fearful messages can be of service to us so long as we have the freedom to pay attention to them or not. The ego's way of fixing the ego would be to make a list of techniques for "working on ourselves," becoming less "this" and more "that." It would be a tough slog—at least sometimes it has been for me—but it is so *boring!*

> ## Maybe You CAN Do It All
>
> During graduate school my department planned an overnight camping trip at an observatory. While I love camping and wanted to connect with fellow students, I lived more than two hours away from school and assumed I wouldn't be able to make it. A week before the trip, a friend pressed me on the issue, so I asked where the observatory was. It turned out to be just a few minutes from my house! Unfortunately, my daughter had a birthday party to attend the next morning, so I checked her invitation to see where the party was—only to find that it was being held at the same campsite! I attended both events without even having to move my car.

You see, your authentic self is *rad*. Living authentically isn't some dry practice of analysis and critique (although sometimes those things are needed to cut through the layers of crap sitting on top). Your authentic self is someone that most people in your life would adore if they saw it. It's composed of the things you truly love, the personality traits you genuinely embody. It's who you are when you're not thinking about what other people are thinking about you. *Being authentic is just being you, undiluted.* Since we are all seeking to express our authentic selves, we tend to be attracted to others who are expressing their authentic selves. That's why we love rock stars, actors, or authors. The more we express our authenticity, the more like rock stars we become.

Is your authentic self dark, disgusting, or dangerous? If you brought your authentic self to work, would it be inappropriate or disruptive? I don't believe so. I believe that as you remove layers of inauthenticity you more clearly understand what is essential. It is not an endless descent into repressed Freudian sexual desires, or at least there is no need to dig that far to find your authentic self; there is a wonderfully vibrant personality at your core. When you find this authentic persona, you can stop digging and just *be*. Rather than thinking about how to fit in, you clearly understand what really matters to you, and from that understanding flows a clear sense of purpose. Your authentic self can appreciate social rules and function normally in the world, going along with things for the sake of pursuing your purpose. Sure, the world isn't perfect, but your authentic self doesn't mind compromising on the little things. Your authentic self is not going to make a scene about every little thing it sees wrong in the world, but it will make waves where it counts. Your authentic self knows what it believes and is not afraid to speak up for it.

This is why the world needs your authenticity. This is why, as leaders in business, public service, education, academia, and every other field, we all need to uncover our layers of reactive programming from wherever we got them and choose instead to be our authentic selves. Once our ego habits are not running the show, our authentic style comes through naturally, and our capacity to create incredible experiences and solve important problems blossoms. Rather than being disruptive or infantile, our authentic selves are appreciative of others and grateful for the little things in life. Gratitude emerges because our authentic selves focus on what really matters to us—maybe making a difference in someone's life or having a real connection with a group of people—and any annoyance with missing a bus simply drops off the radar.

Being authentic doesn't mean becoming a better listener or better at dealing with interpersonal issues. It's not a skillset you need to develop or some person you could become if you worked hard enough to perfect yourself. It's just you without the overlays. The path to accessing your authentic self will be unique to you—that's its defining feature. Yet even though each path is unique, there are some practices that have helped me wade through the layers of rubble to identify what is really me underneath. I will describe a few of them below.

Make Space for What You Authentically Want

It's important to distinguish our cravings from those things that call deeply to us. When someone asks if you want a doughnut, maybe it's a good thing to have restraint (at least, most of the time). But what if somebody asks, "Do you want to bring your kids to the water slides with us on Saturday?" The response that comes to mind may be, "That sounds nice, but I've scheduled Saturday to organize my files for my tax preparer"; but that may not be what you authentically want.

Our authentic self knows what it wants. Yet many of us have been conditioned to be responsible and choose things we don't want instead of those that call deeply to us. That's the difference between the doughnut and the water slides. The day at the water slides with the kids will create an incredible experience that you may remember forever, but by tomorrow morning you will have forgotten about the doughnut. Choosing to say yes to what our authentic self wants makes us more powerful and effective elsewhere in our lives because these experiences recharge us.

But taxes are also important, you say. With flow it's not either–or. Allow space for synchronicity to guide you. By choosing to dive into an authentically rewarding experience, you are aligning with circumstances and trusting that circumstances will align with you. Maybe you'll find that you have an unexpected cancellation at work the next day, giving you have plenty of time to deal with your taxes then. Maybe you'll find that, in the end, the detailed calculations you were doing for your taxes were not necessary. How frustrating it would be to find that out after you passed up the chance to take your kids to the water slides! The trick is that you can't know in advance. With flow, you let go of the worry about being in control, learn to listen to your authentic self, and trust that a needed opportunity will emerge to solve your problem, whichever path you choose.

How do you listen to what your authentic self really wants? Combs and Holland say, "When we allow ... play, the Trickster brings insights about our unconscious hopes, fears, and passions. In doing so he frees us from the compulsion to act out motives we do not understand. The acting out of unconscious motives is the opposite of true play; rather, it means possession by an archetype and therefore the overwhelming of the ego, a dangerous form of psychological blindness."[100] If we consistently choose to take care of our responsibilities first and miss out on the joy of our lives, we are not in flow. This is what Combs and Holland call "acting out unconscious motives," and although we may feel very justified in such decisions, because we are not in flow we are likely to cause pain or regret for ourselves or others. Yet the same is true if we consistently shirk our responsibilities to follow what's most enjoyable in the moment. That is not living in flow, either. It is when we let ourselves *play* that our inner knowing will tell us which choice is right in each unique moment.

"Play" doesn't necessarily mean being silly and spontaneous. One way to find what we authentically want—to distinguish doughnuts from water slides from taxes—is to spend time with distractions turned off. Quiet, reflective time allows the preprogrammed reactions to fall away, and we can begin to hear what is calling to us. However, this does not need to take place on a meditation pillow. We can practice this during our commute (studies have shown the benefits of using our commute time

> ### Flowing into a Fresh Start
>
> "After my mother died, I found myself at a loss. I had left my job two years earlier to care for her, and I found myself asking God, 'Now what?' I wanted something new. I wandered into a bookstore and ran into an old acquaintance who was the director of a women's alcohol and drug rehab center. I caught her up on my life, and she said, 'Come work for me!' I did, and I found that I had natural counseling skills. That job led me to return to college to get a counseling certification and turned into a career I loved." (Story contributed anonymously)

consciously[101]) or at our kids' sports practice. To do this, first turn off the radio or the cell phone, walk a few feet away from other people, and, most of all, keep your mind from straying to pointless topics. Gradually allow your recent experiences to come into focus. Reflect on what you've said and done, and notice any automatic reactions you've had to things. Maybe you said "no" when you wanted to say "yes," or vice versa. These can be clues to where your authentic self is being hidden, areas of your life where you are not making space for what you authentically want. Just the act of concentrating on the meaningful questions in your life can open up hidden possibilities.

The ability to do this is called "trait self-control," and studies indicate[102] that it may lead to greater happiness and life satisfaction for those who practice it. One day in 2009 I was driving on a country road on my commute home, and I can still picture exactly what I was looking at when I had the insight that led to my understanding of retroactive event determination, a major factor in the development of my ideas on synchronicity (discussed in the next chapter). As I was looking at the landscape around me, I suddenly recognized that the things I saw at that moment—the hills, the grass, the trees—were all I could claim to be certain about; anything in any other part of the world was inaccessible to me.

This insight dramatically influenced the unfolding of my subsequent research. I don't think it would have come to me if I'd had the radio on or if I were obsessing about something my boss had said to me at work that day. It happened because I was using my commute as distraction-free time to concentrate on physics. Of course, you might find such an activity excruciatingly boring, but it is a delightful experience for me. What would benefit from *your* careful, focused attention? What questions can you ponder in your in-between moments if you carve out a little more personal thinking space around yourself?

Daniel Pink, author of *When: The Scientific Secrets of Perfect Timing*, emphasizes the importance of these in-between moments. In an interview for *GQ* he said, "My [previous] view was that amateurs took breaks and professionals didn't. That's just diametrically, 100 percent erroneous. Professionals take breaks, amateurs don't take breaks."[103] He emphasizes that a break doesn't mean "spend[ing] the whole time answering text messages from my boss or looking at my Instagram feed." By choosing to make space for meaningful thinking, you are likely to notice things you may have missed. You may suddenly recognize a chance you have in your life to speak out about something important or to create a cool situation. You may have some insight regarding how to approach an issue at work or in your family that will breathe new life into it.

Carving out time to think deeply about any meaningful topic can release a powerful indwelling creativity. You may walk away from this thoughtful downtime with a refreshed perspective and renewed clarity on what you really want. This leads to the next practice: once you know what you really want, use every chance you have to dive into the flow and create it.

Turn 'Freedom to Check Out' into 'Freedom to Create'

Why do we want to check out of life sometimes? Why do we love vacations, sports, movies, and drinking? Such activities can be wonderful things, but I suspect that the current cultural obsession with leisure is not a fundamental human trait; it arises in contrast to the exhausting world of work and effort we have created. When we are in flow, however, work is relaxing. When we make space in our lives for what authentically fulfills us, we are naturally recharged. Activities that might technically be labeled as work feel quite a lot like play.

Computer scientist Margaret Butler describes her feelings about building computers at Argonne National Laboratory:

I may work ... hard ... out of ambition or a desire to make money. But unless I also enjoy the task, my mind is not fully concentrated. My attention keeps shifting to the clock, to daydreams of better things to do, to resenting the job and wishing it was over. This kind of split attention, of half-hearted involvement, is incompatible with creativity.[104]

Csikszentmihalyi has more to say about the nature of work in modern life based on the group of professional women he studied:

Most women who work at clerical, service, and even managerial occupations tend to think of their outside job as something they want to do rather than something they have to do....

Many of them feel that whatever happens on the job is not that important—and thus, paradoxically, they can enjoy it more.[105]

The golden boundary between work and play seems to have a lot to do with creativity and flow. When we want to check out at the end of the day, it's because we have been out of the flow. By contrast, when we are immersed in a task and notice the meaningful coincidences and connections between the various aspects of our work, we come alive in the same way a third-grader comes alive on the playground. The third-grader is so enthralled with the creativity of the game they are playing that time falls away and recess feels like it ends too soon. Of course, for some kids it is not recess that brings about this experience but rather art class, science, math, drama, history, reading, or creative writing. Because kids are often sheltered from the need for a particular outcome, they are free to dive into the flow of experience.

If we don't feel free in our work, we will look for opportunities to check out. When our normal experience is one of flow, there is nothing to check out from. If we felt totally free to be creative, I suspect we could each come up with many interesting things to dive into. Of course, if we are not in the habit of feeling free to create, we might not be able to think of anything we really want to do. We might not realize how many things we would authentically enjoy doing because we simply haven't seen them as options. It's easier to seek simple ways to check out because they don't require any effort or momentum. Creativity, on the other hand, is developed through the investment of energy over the long term. We have to build momentum in our creative endeavors in order to bring about synchronicity; thus, the payoff is slow and gradual.

In order to gain more freedom to create in your life and kickstart the flow, try the following approach. After reading and pondering the previous section of this book, you might have a clear sense of authentically pursuing something in your life. Put yourself in a room or other space that has some variety or elements of interest related to your project. Pause whatever you are doing, step back from your tasks, and look around. Do anything spontaneous that gets you thinking about your passion, whether it's opening a book, moving a piece of furniture, or writing a few thoughts down on paper or on the computer. Allow whatever comes to evolve and guide you to other activities, wherever your curiosity leads. The only constraint is to limit your curiosity to the project at hand, rather than going off down unrelated tangents. Stay clear on what the project entails, and stay on that path. In this way you walk a path between free-for-all flowiness and ultrafocused rigidity.

The beauty of this approach is that you can apply it to employment-related activities as well. If your task is to gather tax data for end-of-the-year accounting, use flow to approach the work in a creative space rather than rushing through it laboriously. This might allow you to feel a sense of ownership and creativity in the process. Maybe you will come up with a new way of organizing or gathering the information, moving the process forward by leaps and bounds for next year. When you're done, you won't feel the need to check out. Instead, you'll be wondering where else in your job you can apply that same creative energy.

Synchronicity Happens through Us

Authenticity is not just about expressing ourselves more clearly. Even as we get better at being authentic, we find that we are part of something bigger than ourselves. We might feel like we are pawns in a chess game that is playing through us, as I felt during my spontaneous gig at the Cheeseboard. I did what I did as an authentic expression of myself, but it ended up having a meaningful impact on my nephew and daughter. When we see through the heart, we act simultaneously for ourselves and for others. The beneficiary of our actions is the entire web of life we are part of. We are the vessels through which love expresses itself in the world, and synchronicity is a way in which love's expression is orchestrated.

Western society gives many of us the clear message to focus on our own lives. We are asked what school we will attend and what we will do after we graduate. We are supposed to choose a career and buy a house to build a stable financial future. These expectations convey the underlying message that personal security and stability make for a successful life.

Is it enough to be stable and successful? If everybody is stable and successful, will our culture thrive? Csikszentmihalyi says no: "The quality of life does not depend directly on what others think of us or on what we own. The bottom line is, rather, how we feel about ourselves and about what happens to us."[106] He references data that indicate "a mild correlation between wealth and well-being," but he points to research done by Diener and colleagues stating that statistically very wealthy people are only moderately more happy than those of average wealth.[107]

Exclusively following the path of stability and outward success leaves out an important aspect of who we are: creative creatures who can employ our skills and talents in the service of something bigger than ourselves. It is clear that many people want to act on this aspect of themselves, as demonstrated by the increasing number of financially successful individuals who care passionately about the well-being of others. Living in flow helps us bring out our creative selves in a way that doesn't diminish our own likelihood of success but rather enhances the quality of that success, for us and for others. We know from research that there is a point at which acquiring more wealth does not contribute to a greater sense of well-being. It seems logical that it may be precisely this transitional point at which an ability to live in flow becomes the better predictor of happiness.

> ## Following the Flow of a Career Path
>
> "As a teenager I would check out the boys who worked as baggers at the local grocery store. One day none of the boys were working, so I asked the store manager where they all were. He must have misunderstood me, because he handed me a job application. I decided to fill it out anyway and ended up working there for ten years until it closed. Years later I had just moved back to town and stopped in at a sister store. My friend said they were having a grand reopening and handed me an application. I received a call back from one of the managers I had worked with as a kid, and I was hired immediately." (Story contributed by Susie Hicks)

A beautiful aspect of this approach is that when we align with circumstances—whether by accepting our current circumstances or fighting for a new circumstance we wish to experience—we find paths forward that serve everybody. We don't have to choose personal success at the expense of the prosperity of others. By living in flow, we can have both.

I think we can use meaningful history selection for personal gain, but the more we see it acting in our lives, the more we understand the interconnectedness of all things. Our worldview shifts and matures into a new paradigm that integrates competition and independence with collaboration and interdependence. Using synchronicity as a framework for our lives makes us wise, because it opens us up to our own authentic desires and vulnerabilities. It also makes us humble because it helps us understand our own shortcomings. I suspect the cosmos responds not to our ego's desires but to our authentic nature. I also suspect that our authentic nature (if we can access it) is based on a healthy balance of caring for oneself and caring for others. We naturally experience joy when we see others experiencing success, so long as we trust that we are also getting what we need. By living in flow and noticing synchronicities, we can lessen the feeling that our life is lacking—lacking money, success, opportunity, appreciation—and then we naturally want to see others thrive.

Some readers might think, "Of course I feel that being altruistic is super-important. I'm already that way." Altruism can also be an obstacle to authenticity and synchronicity. Maybe we don't want to speak up in a meeting because we think somebody else might have something more important to say. Or we don't want to talk about a project we are working on because it feels like self-promotion. But synchronicity works *through* us. It is our comments that may be a tipping point in somebody else's world. It is our actions that may brighten somebody else's day. Many of us tend to undervalue our impact on the world, thinking that we are just slinking along through life unnoticed. I suspect the truth is the opposite: in meaningful history selection, every action counts. How does the world change, except through the efforts of people like you and me? The world needs our gifts. So there is a communal benefit to developing our individual gifts and expressing them. If we are unwilling to share our gifts, potentially meaningful branches of the tree will never happen.

When we live in authentic flow—listen, open, reflect, release, and act—we are more likely to say just the thing that somebody else needs to hear or to come up with a plan that perfectly aligns with the needs of another person. We may not realize the ways in which our actions are affecting the larger flow, but that's okay. The idea is that when we are in authentic flow, we can trust that our actions are having a meaningful impact.

I don't think living authentically means we should drop out of the system and give up any concern with materialism. I have a nice home, many material possessions, and a family I love deeply. But these things aren't the end result of my life; rather, they support me in doing the task the world really needs from me, which is to live purposefully and authentically. Being authentic does not mean rejecting everything that isn't fully authentic. If you are a lawyer, I am not saying you should abandon law just because you realize you have, say, a desire to serve underprivileged people. The greatest challenge may be developing more authenticity within your present life. An opportunity may arise to take on a case that you feel called to pursue but that risks your reputation with your peers. *This* is the turning point of authenticity. Becoming more authentic may not involve an overhaul of your master plan; it can be about becoming more fully alive in your daily decisions. What choices have you made based on what society values? How can you be more authentic within the path you are already on?

Authenticity in Public

Authenticity can be a powerful guide for living in flow. When we are authentic, we are committed to following life where it leads us, and this includes speaking from our hearts in ways that might have previously felt uncomfortable. We have already talked about how flow can enhance the quality of our relationships and can lead us to new relationships that are meaningful. Living in flow can also lead us to be more authentic with our values in public.

I once experienced a synchronicity that illustrates the power of speaking up in public. First I had a dream in which I was encouraged to write something that gets noticed. I spent some weeks trying to write bits and pieces, but they were not very good. Then one afternoon my daughter and I were at a grocery store together. Standing in line at the register, I noticed a magazine that was filled with hate speech. I couldn't quite focus on paying for my groceries, and realized I was disturbed by what I had seen, so I decided I had to speak up. When the manager arrived to speak with me, I started by empathizing with the challenges he must face in his role, and I assured him that my concerns did not reflect on him personally, but I wanted his help in addressing them. Once I told him why it was upsetting to find this type of hate speech on the shelves in my neighborhood store, at an eye level where my young daughter could read it, he replied, "If that's the way you feel, we'll take them down." He proceeded to remove them from every register.

I ended up writing an article about the incident, and it went viral. My dream had given me a motivating clue, and then the flow of synchronicity led me to a unique opportunity to make a difference. In order to grab that opportunity, I had to speak up in an uncomfortable way in public. I did wrestle with the issue of freedom of speech. The issue was not whether the magazine had a right to be there; the issue was that the store's choices to feature the material—which were apparently not made consciously—were at odds with basic decency standards of their clientele.

I have paid particular attention to the feelings that come up for me in public settings, and in so doing I have found a strange paradox that contains a hidden gem. Most of the time we feel shy about speaking up in public; when we are waiting in line at the movies, we are not likely to say out loud to the whole line, "Aren't we so excited about this movie?" Yet we speak up quite readily when we feel anger or danger. In emergencies people break the spell of isolation and are motivated to work together. There is a

Authenticity as Flow

layer of resistance called the "bystander effect." This effect keeps us from engaging in public on a daily basis, but the hidden gem is that we all know how to do it in certain situations. This has led me to a question: What if love and positivity could empower us to speak up for what is right just as anger and fear so easily do?

The Heroic Imagination Project is a nonprofit organization that does fascinating work in training everyday people to see themselves in the role of everyday heroes.[108] They help people develop a sense of self-efficacy and responsibility in situations where others may feel powerless to make a difference. This requires drawing on the power of our positive emotions just as willingly as we might, at other times, draw on the power of our negative emotions.

As a parent, for instance, I am ashamed to admit how willing I am to act like a jerk in public when I am really angry about something, doing things I would normally never do. As I mentioned in the discussion on boldness in chapter 4, it is easy for me to speak up boldly in public if my daughter accidentally spills her soda on me after I have asked her to stop horsing around. The hormones released in response to anger make that quite easy! But if a great song is playing in the background, it is not so easy to spontaneously start dancing in public, even if I feel the urge. Why not act with the same courage when we feel joy? Synchronicity and flow push us to do so.

As I have matured as a performer, I have developed a greater admiration and respect for professional performers. I have come to admire those who have a willingness to break through resistance and engage people authentically on a public level. Most importantly, they do so from a place of love and creativity rather than anger and fear.

For example, the song "Brave" by singer-songwriter Sara Bareilles addresses this very topic. The song's video includes footage of everyday people dancing and singing in public. To make this video, Bareilles had to feel comfortable standing in a public square, talking to strangers, and asking them to break their veneer, as she was doing. In my experience, this takes a tremendous amount of courage. Those who have broken through their own self-consciousness in this way can teach us about stepping out in public from a place of love.

Why is it difficult to speak up in public? An airport experience provided some insight into this issue for me. I find that standing in the security line is a strange conspiracy of silence. There is nothing really to say to a group of strangers in that situation, yet we are all sharing a similar experience, so the silence is palpable. It seems like there is a layer of restraint, a boundary of silence, that is hard to breach. On this particular day, a woman appeared whose flight was leaving in fifteen minutes. This was her second time through the line, and it seemed the natural thing to let her in. Suddenly, I felt a sense of common purpose with her. She was understandably hesitant to speak up on her own behalf, but as her ally I could speak up without feeling selfish. I asked if any of the folks in line would be willing to let her move ahead to the front of the line. In this case, everyone agreed. She quickly made it through security and successfully caught her flight.

The amazing thing was the shift in energy after this event. Once there was a shared purpose, a sense of trust developed among those of us in line, and natural, heartfelt conversation ensued. We could all agree on this simple ethic—helping a person in distress—and that was all it took to break the veneer of silence. We were able to easily converse as a collective about having been in that situation before, making jokes or finding empathy. I had a genuine connection with the person in front of me and behind me, and after the security gate we exchanged business cards.

Before the person in need appeared, it felt very difficult to break through the silence. Once we had a common purpose, I had an easier time speaking up. And once the silence had been breached with that common purpose, we became a group of friends rather than a group of strangers.

It seems to me that many of the things that go wrong in the world can be addressed by bringing them up in public. For instance, how often are we sold a product in excessive plastic packaging? Yet we are becoming aware that plastic waste is leading to vast areas of lifeless ocean around the planet. The first important step is to notice when it happens: somebody serves me a plastic to-go box or a ridiculous wrapper that I simply don't need. Once we have realized what happened, what are we to do? If we, the actual human beings in the store, got together and had a genuine conversation about it, we might begin to address the issue. At the very least, public acknowledgment of the depth of the problem is a necessary step toward eventually solving it.

How do we do publicly acknowledge the problem? We must find common purpose and break the layer of silence. If we don't, we let our silence shape our world. What's amazing to me is that we *all* have common purpose; we just have to find it. No matter where you are and who you are with, you share common experiences, feelings, and purposes. Having the courage to breach the silence in public seems to be about the challenge of finding common purpose so we can establish mutual trust.

How do we know when it's the right time to breach the silence? Once again we have to return to our inner sense of knowing. By knowing ourselves and our own values, we can tell whether a circumstance is meaningful and important to listen to. The cosmos may send clues, like the woman who needed help getting to her flight. It is up to us to decide whether this event carries some form of guidance for us, something useful we can learn from or give back to the experience.

Although the opinions of strangers can hold us back, these relationships are also a key part of making the world a better place. At the end of the day we can only move forward on bigger issues by finding common ground and making progress collectively. And it is not just strangers in public whom we can respond to with a sense of constraint. Even within familiar relationships we may feel reticent to talk about certain things or do something spontaneous that we want to do. Relationships can sometimes hold us back, but a commitment to authenticity can help us break free. By noticing the feelings of hesitancy but choosing to act from a place of courage, we may open up to a greater depth of relationship.

Synchronicity and flow can be valuable assets in helping us have the courage to be authentic in public. When we feel the urge to do something that might make us feel vulnerable or exposed, I suspect that the process of meaningful history selection makes it more likely that we will experience coincidental situations that make our bravery pay off. Understanding this process has made me more confident in taking risks like this; I recognize how many of the meaningful opportunities in my life have only happened because of previous risks I took, and I have faith that the same thing will happen again. This gives me the courage to speak up authentically in public, however that looks in the moment.

Developing a sense of authenticity while in public can help us find our sense of purpose as well. When we are free to be ourselves, we are free to follow the flow to discover how we may be of service in any situation. When we trust in the response of the cosmos to our choices, we become more willing to stand up for what we feel is right. This is a tool in our belt as we look to the future. We are no longer looking for solutions to our problems (the content); rather, we strive for a new attitude toward ourselves (the context), which empowers us to be the change we wish to see in the world.

In Part I we covered the basic ideas of flow and synchronicity from a variety of perspectives, including the ways in which flow and synchronicity can be cultivated in your everyday experience. In Part II you are invited to take a look at the fundamental physics that I suspect underlies these ideas. In chapter 7, "Exploring the Foundations," I offer a brief introduction to some of the relevant aspects of modern physics, including special relativity and quantum mechanics, as well as some of the new ideas needed to support the concept of meaningful history selection. In chapter 8, "Meaningful History Selection," we examine again how this process leads to the experience of synchronicity, but from a more rigorous point of view. In chapter 9, "You Are a Spark!," we reaffirm the importance of living in flow and how I suspect your choices shape your world.

PART II

7

EXPLORING THE FOUNDATIONS

I think it's important for people to feel connected to science. Unfortunately, science and math are often taught in schools in a way that leaves a portion of the population feeling like they just aren't science people. Also, advanced physics becomes very abstract, and physicists become so specialized that they feel little impetus to communicate their ideas in a nontechnical way.

Yet in an age where nearly everything we do is influenced by advanced technology, I hope nonscientists can learn to embrace the "sciencey" part of themselves and that scientists can remember the importance of presenting their ideas in a digestible manner. I hope that concepts such as meaningful history selection, in particular, influence how people think about their everyday lives.

This chapter and the next are intended to serve as an accessible background for the proposed science that leads to my conclusions on synchronicity and flow. I hope you will find it enjoyable while resting easy with any concepts that don't connect for you right away. In using the left brain and learning new technical ideas, I find it helpful to read the first time and just absorb whatever I absorb. If you are called to return to it a second time, more of the ideas will likely make more sense.

Experience Requires an Observer

My research is in a field of physics called "the foundations of quantum mechanics," or "quantum foundations." This is the study of what quantum mechanics "really means." Quantum foundations is to mainstream physics as the glee club in the TV series *Glee* was to the rest of the school: valid enough to have a spot on the school schedule, but still considered a waste of time by many. Why is this? The answer is steeped in history.

In the years between the world wars, physics made great strides in developing quantum mechanics. The people who did this were, in my view, like the great statesmen who wrote the U.S. Constitution: visionaries who thought through the deeper consequences of what they were doing. Their research uncovered fundamental connections between the dry, objective field of classical physics and the juicy and subjective world of human experience. The new theory of quantum mechanics implied that every measurement of an object must also take into consideration the one doing the measuring, i.e., the observer. The observer wasn't just an interpretational concept to philosophize about; it showed up in the mathematics as well. Let's examine briefly how you, the observer, enter into quantum physics.

According to standard quantum mechanics, there are two ways things can change. The first kind of change is a predictable process that tells us how any object in a known circumstance evolves to a new circumstance. This type of change is called "unitary."

The second kind of change occurs when objects interact with other objects. The interesting thing about this kind of change is that a given interaction might have a lot of possible yet unpredictable outcomes. When two billiard balls collide, they might go right where you intended, or they might fly away all catawampus. Analogously, when you walk through the grocery store and pass an old friend, you might not see her and just pass by, or you might recognize her and stop to talk. Any interaction can generally lead to many possible outcomes, depending on the how the interaction goes. In quantum mechanics, we use a mathematical tool called a matrix to understand how the objects change. Each matrix has specific solutions to it that tell us what outcomes can result from the interaction.

Where does the observer come in? Imagine yourself as a traveler standing in a train station in Florence, Italy. Each train represents a different measurement you could choose to perform. One train has the possible outcomes of Rome, Naples, and Messina. Another train has the possible outcomes of Pisa, Genoa, and Milan. A third train has the possibilities of Bologna, Venice, and Trieste. The object being measured is Italy itself, one outcome at a time. You, a human observer, can choose which train to take, or which matrix to apply. The important point to note is that the human observer's choice of measurement determines the available outcomes. You walk away with a particular description of Italy based on the train you decided to take (e.g., the Bologna-Venice-Trieste train) and the particular outcome (e.g., Venice) you received. Venice is a different experience of Italy from Rome, yet they are both valid measurements of aspects of Italy. Just because you saw Venice, however, doesn't mean you know everything about what Italy "really" is like. Your choice of train determined what you learned about Italy.

Your motion through the countryside, looking out the window and enjoying the breeze, is guided by the train and corresponds to the predictable unitary aspect of quantum evolution. The decision of which train you take, however, is up to you, with far-reaching consequences for where you go. Once you choose a train, there is no rule for predicting with certainty which of the three cities you'll end up in. Only the likelihood of each city can be known.

Therefore, the result of any scientific measurement in quantum mechanics depends on both the thing being measured and the one doing the measuring. Quantum mechanics irreversibly messed with the minds of materialist physicists by bringing their own conscious experiences and choices into their models of objective reality.

Classical versus Quantum

"Classical" physics simply refers to our understanding of science before the development of quantum mechanics. This is how we viewed things until around the year 1900, based on the work of Galileo Galilei, Sir Isaac Newton, James Clerk Maxwell, Lord Kelvin, and many others. "Quantum" physics (or, equivalently, "quantum mechanics") is a set of ideas that has its own totally novel approach to describing reality. The two approaches give the same answers in many cases, such as calculating orbits of planets and the motion of automobiles, but they give very different answers in other cases, such as electronics and cryogenics (the science of very cold temperatures). Although quantum mechanics has some incomplete aspects, its basic principles are extremely well-tested, and we are more confident than ever that it has been a massive step forward in our understanding of the physical world.

I think the most important contribution of quantum mechanics to human understanding is the understanding of potentialities as real things. In classical physics, things are described as either existing or not existing. A baseball is either in your glove or not in your glove, and this is the difference between the base runner being out or being safe. In quantum mechanics, however, we don't deal with things, but rather with *properties* of things, and those properties are described on three levels: nonexistence, existence, and possibilities or potentialities.

For instance, in quantum mechanics, if we want to describe an electron (the tiny subatomic particle that conducts electricity), first we will say that it either doesn't exist or it exists. If it exists, it may be known to be spinning in one of two directions, either "up" or "down." An electron may also be *unmeasured,* meaning it has a potentiality to spin either "up" or "down." By "unmeasured" I mean that you have not interacted with it in some way, such as by looking, listening, feeling, using a tool like an amplifier, a microscope, a pair of tongs, etc. The key point is that if it's unmeasured, then its spin property is a combination of two distinct possibilities. Because you haven't examined it for yourself, it doesn't definitely "exist" in either condition. Yet it does "exist" in the sense that if you *do* measure it, you *will* find it there in one form or another, spinning either up or down.

The idea is that we can't just think in terms of definite, concrete objects in the world. We have to think of each thing as a set of potentialities until we look directly at it for ourselves and are therefore sure we know what its condition is. This may seem abstract or intellectual, but these possibilities are "real" enough—they can be manipulated and can have measurable effects on the world. Semiconductor technology, such as the kind used in mobile phones and computers, couldn't function without this intermediate level of possibilities.

What do I mean by this? Can't we just say, "Of course there is a world of possibilities. My future is full of unknown possibilities! That's not quantum mechanics, that's just normal unpredictable life, right?" Well, hold on a minute. Quantum mechanics doesn't just say the *future* is unknown. It says the same about the present and even the past. Quantum mechanics says that anything that hasn't been observed is unknown—and not just "unknown" as in "somebody knows, just not me." In quantum mechanics, things that haven't been observed or witnessed don't have definite properties at all. They instead exist in a condition known as "quantum superposition," which means that when a measurement occurs they will exist in one of the listed states. So if someone tells you that quantum mechanics allows, say, an electron to be in two places at once, they're being lazy. The correct way to say it is that there is a *property* of the electron we call its *location,* and the electron doesn't have a definite number associated with that property. It is not in two places at once; rather, there are two potential places where it could be found to exist—*if* we measured it.

So quantum mechanics can be summarized as *the study of what the world is doing when we are not looking.* Physicist David Bohm put it this way: "In quantum theory it has no meaning to discuss the actual state of the system apart from the whole set of experimental conditions which are essential to actualize this state."[109] In other words, we can't separate the answers we get from the questions we ask to obtain them.

It seems to me that we make all sorts of assumptions about what the world is doing when we are not looking. In Evita's case, while she was applying to graduate school, she may have assumed the department had carefully read her application and rejected it, or she might assume that her application got lost in the shuffle and that she got a bad deal. Generally, her assumptions could be wrong or right. But either way, she is still assuming that *something certain* is happening, and if she could only learn about it she would know. I think quantum mechanics says no. The properties of things that we aren't observing simply don't have a definite value yet.[110] Evita's application status doesn't have a definite history or a definite set of events that actually happened. It remains just a collection of possibilities.

Now, there is a vigorous debate about the scope of this fact. According to some ways of interpreting the theory of quantum mechanics, this only applies to invisible, microscopic systems and thus shouldn't really be of interest to most of us (unless we are theoretical physicists). However, this is why people think of quantum foundations as a waste of time: the mathematics seems to imply that quantum mechanics *should* apply to all objects, but we just can't seem to intuitively grasp how that squares with the world we actually experience. Many people shrug and move on to more tangible problems. Yet there are many reputable interpretations of quantum mechanics that do not make this assumption, and the situation is far from solved in either direction.[111]

In my work, I don't assume that I should expect to directly observe quantum mechanics acting in the real world, because quantum mechanics is the study of what things are doing precisely when we are *not* looking. So I can't simply trust my perceptions to tell me whether quantum mechanics is there. I suspect that, just as with microscopic electrons, the properties of the world right outside my door remain uncertain. Will a package I've been waiting for arrive when expected, or will it show up a week late? Will a salesperson arrive at my door and ring the bell right when I lie down to take a nap? I suspect that these macroscopic events are also properties of the universe described by quantum possibilities. In other words, although the baseball will always be either in the glove or out of the glove *from the perspective of the umpire,* both the umpire and the ball remain as a collection of possibilities *from the perspective of someone not in the stadium.* This perspective-driven view is called relationality, and I will try to make the case for it by the end of the chapter. It has been one of the subjects of my research in quantum foundations.[112]

Probabilities and Measurement

To understand the way a physicist thinks about this, there's a term we need to learn. Any property of an object, such as its position, is described by a numerical value called an "amplitude." The amplitude for each of the possible positions of the electron tells us how likely that position is to actually occur, if we look. When we do look, only one of the possible positions will become the *actual* position of the electron. Each potential position of the electron has its own amplitude and therefore its own particular probability of becoming true. This is very different from the electron actually being in two places at once. It has two possible states in the intermediate level of potentialities, but only one of them can become actualized in the world.

Physicists have discovered that when we square the amplitude, we get the "probability" that we will measure the electron in that place. So there are three levels to existence: nonexistence, where an object simply doesn't exist; possibilities, in which each possible value for the property of an object has an amplitude associated with it; and existence, where the property has a definite value and that definite value has a certain probability of occurring. So while an amplitude describes possibilities that *could occur*, a probability describes outcomes that *really do occur*.

I go to the trouble of describing the math here simply to emphasize this point: we should stop thinking of a black-and-white world where things either exist or don't exist. Science is clear (although not all scientists interpret the facts in the same way) that we need to think also of an intermediate layer of possibilities described by amplitudes. This is a crucial feature of the cosmos that will lead eventually to what I call the "responsiveness" of the cosmos, from which come synchronistic experiences which lead to flow.

There is one final important aspect of quantum mechanics to mention, which brings us back to you, the observer doing the measurement. Imagine again that you have an electron. The location of the electron is not fixed if you have not yet looked

> ## Discovering Matter Waves
>
> The "double slit" experiment is the quintessential experiment in quantum mechanics, showing that light behaves both like a particle and like a wave. Serendipity led Clinton Davisson and Lester Germer to discover that electrons demonstrate this property too. They were shooting electrons into nickel metal when the vacuum tube unexpectedly exploded due to an air leak. In order to redo the experiment they had to use a high-heat oven to remove oxidation from the nickel. The oven unexpectedly created some larger crystals on the otherwise smooth surface. The next time they ran the experiment, these crystals acted like a double slit, creating a visible interference pattern that led to the discovery of matter waves.[113]

at it. Then, when you measure the electron, it jumps into a fixed location. Yet you can't simply say you are finding out about the true nature of the electron, because the location that the electron jumps into can only be one of the possibilities that your measurement allows. The electron's properties depend on the type of measurement you do—or, put another way, on the questions you ask.

You might see in this fact the connection to a responsive cosmos. You can't ascribe the properties of any object simply to the object itself. Rather, the properties that you observe—of where and when people appear, for instance—are influenced by the sorts of questions you are asking. In case you feel too comfortable too soon with this idea, try to remember this is more than just a glass-half-full or half-empty debate, more than a question of whether you see things optimistically or pessimistically. I am suggesting that the very events that show up in your life may be a response to the sorts of actions you are taking. It's about more than how you interpret events; it's about whether a specific event happens at all. Keeping in mind that this interpretation is far from certain and not agreed upon by most physicists, I will do my best to justify it shortly.

Carl Jung, the twentieth-century psychologist known for coining the term "synchronicity" and for his work on the collective unconscious and the symbols known as "archetypes," explains this dilemma well:

Experiment consists in asking a definite question which excludes as far as possible anything disturbing and irrelevant. It makes conditions, imposes them on Nature, and in this way forces her to give an answer to a question devised by man. She is prevented from answering out of the fullness of her possibilities since these possibilities are restricted as far as practicable. For this purpose there is created in the laboratory a situation which is artificially restricted to the question and which compels nature to give an unequivocal answer. The workings of nature in her unrestricted wholeness are completely excluded. If we want to know what these workings are, we need a method of inquiry which imposes the fewest possible conditions, or if possible no conditions at all, and then leaves nature to answer out of her fullness.[114]

The discoverers of quantum mechanics were aware of some of the philosophical and existential challenges posed by quantum mechanics. Werner Heisenberg said, "The concept of the probability function does not allow a description of what happens between two observations. Any attempt to find such a description would lead to contradictions; this must mean that the term 'happens' is restricted to the observation."[115] In other words, we cannot talk about what exists or doesn't exist apart from the interaction a thing has with the rest of the universe. It is only a short leap from there to see that we can only speak of properties of things, not of the things themselves.

Physicist Wolfgang Pauli appears to have been quite interested in the phenomenon of synchronicity. However, he did not feel that synchronicity was related to the theory of quantum mechanics that he was instrumental in formulating, principally because quantum theory can be reliably replicated, whereas synchronicity seems hopelessly unpredictable. His article "The Influence of Archetypal Ideas on the Scientific Theories of Kepler" indicates that he was deeply influenced by Jung and that he suspected, as I do, that symbolism may underlie both quantum mechanics and the phenomenon of synchronicity:

> *As ordering operators and image-formers in this world of symbolical images, the archetypes thus function as the sought-for bridge between the sense perceptions and ideas and are, accordingly, a necessary presupposition even for evolving a scientific theory of nature.*[116]

With the advent of this set of theories and results in quantum mechanics, together known as the "measurement problem," it seemed that physics had come full circle to reinvent old ideas from ancient traditions.

Shut Up and Calculate!

In the midst of all these new ideas, World War II broke out. The great minds of the quantum revolution were separated. Werner Heisenberg, a German national, was isolated in Nazi Germany during the war. Niels Bohr was in Nazi-occupied Denmark, and the great Albert Einstein, a Jew, escaped Germany just in time and found a home at Princeton University in New Jersey.

Yet even Einstein, who wrote a crucial letter to U.S. president Franklin Roosevelt urging him to develop the atomic bomb to defeat Nazi Germany, was redlined from the project itself and mistrusted by the U.S. government. The world was in survival mode, and practical thinking became more highly valued.

After the war ended, the Cold War with the Soviets began, and this practical focus continued. The rapid growth of the military-industrial complex, along with funding for the many practical applications of physics that had been developed in this period, led quantum physicists to a way of thinking that was later captured in N. David Mermin's pithy phrase "shut up and calculate."[117] In other words, stop thinking about the implications of the work and just calculate the experimental outcomes. This approach was tremendously successful because it led to easily quantifiable results in two forms: data and cash.

Quantum mechanics gave birth to the transistor and, eventually, the solid-state computer. The results of hands-on experiments based on calculations were easy to quantify and attracted military funding, as part of a cycle that dramatically increased the number of physicists being trained in U.S. educational institutions. The development of "quantum field theory," based less on new fundamental principles than on approximations of ever-increasing accuracy, gave rise to the most accurate experimental predictions in all of physics, in the form of particle accelerators like CERN in Switzerland. All of this gave birth to the modern tech and communications industries, whose impact on the global economy is impossible to overestimate.

Quantum foundations, the study of what quantum mechanics really tells us about the world, became unnecessary in the same way it becomes unnecessary to have a live band at your wedding if you have a smartphone. In an ideal world the live band would be valuable to have, but on a practical level, it's not worth the effort and expense. Quantum foundations is a thriving subculture within physics, neglected by many but making progress nonetheless.

I love the field of quantum foundations because I believe that at some point our luck with the "shut up and calculate" paradigm will run out, and a deeper understanding will become necessary. We cannot go on forever building on an incomplete foundation and expect to make progress, and I believe the foundation can be completed.

The Timelessness of Light

How are we to make progress in quantum foundations? My approach starts with examining the properties of light. There is an interesting piece of information in the body of physics literature that I believe hasn't been thoroughly explored. It turns out—and this is an example of research I am pretty confident in but has yet to be affirmed by extensive peer review—that light is timeless.[118]

Many of the greatest historical breakthroughs in physics have come from attempts to understand light, the phenomenon that gives rise to vision, the sun's heat, and so many other essential phenomena. Much of Isaac Newton's research in the seventeenth century centered around light, using prisms to understand the visible color spectrum. James Clerk Maxwell unified the fields of electricity and magnetism in the nineteenth century by showing that these two forms of energy together gave rise to a particle traveling precisely at the speed we knew that light traveled. It was not a big leap to recognize light as a wave of electromagnetic nature. Einstein's special theory of relativity in the twentieth century was inspired by such questions as, "What would I see if I traveled so fast that light couldn't keep up?" Our models of light do become more sophisticated in each generation; yet light is so fundamental that even after many millennia of study our understanding remains incomplete.

To develop an even better understanding of light, let us first consider the mathematics of Einstein's theory of special relativity, which tells us that time and space are pliable. If one were to ride on a light wave, which is the image that Einstein had in his mind as he developed his theory, both time and space would shrink to zero. To say that time and space have shrunk to zero just means that what we normally think of as two separate events, like light being created in the center of a star and then the same light being absorbed by your eye looking up at the heavens, have to be thought of as the *same event*. They couldn't be different events if the space and time between them have shrunk to zero, because two events occurring at precisely the same time and same position must actually be the same event. How else would you distinguish them?

David Bohm was sleuthing on the same trail when he proposed a notion he calls the "implicate order" to describe a holographic, "one-in-all"-ness to the cosmos. He wrote, "In each region of space, the order of a whole illuminated structure is 'enfolded' and 'carried' in the movement of light."[119] In other words, if we look at it right, all of the cosmos resides in every portion of the cosmos. If light is timeless and spaceless, then light would seem to be, in some sense, omnipresent.

How then do we talk about the speed of light? Light has no mass (or what you can think of as "weight"), so it is totally unlike all the physical objects we touch on a daily basis. As a consequence of having no mass, it travels at a fixed speed, creatively named "the speed of light." But if light being emitted from a star and absorbed by your eyes is actually a single timeless event, how can we say that other stars are thousands of light-years away and the light from them is thousands of years old? That has to do with *perspective*. A light particle is a single transference of energy that appears to take time to travel from place to place, *from the perspective of observers like you and me.*

Physicists say that light travels on a "null interval," or a separation of space and time with zero length. This is a fancy way of naming something that is in fact quite mysterious. What it means is that we should stop thinking of light as something that travels smoothly from place to place. For light, space and time are not even defined. Rather, light must only appear when and where it is measured. You and I—existing as we do in space and time—find that light appears here and there just as if it had traveled smoothly between places, but this is just a useful fiction we tell ourselves. It reminds us of a baseball flying through the air from bat to glove, and this intuitive picture comforts us.

A better description for light is that a "possibility for the light particle" (what we call a "wave function") exists "outside" of time and space, and when this possibility interacts with the "possibility of somebody to see it," the observer claims that the light "exists" at the time and place where they see it. Before the person observes the light, it is just a set of possibilities described by amplitudes. If and when the person observes the light, it becomes a real event, causing a flash of brightness on their retina or the feeling of warmth on their skin.

The implications here may be bigger than is initially apparent. These considerations lead to the idea that everything in the world obeys quantum superposition, not just electrons and light. This is why, for instance, in the story about Evita getting into graduate school, I claimed that the situation with her application was not fixed in place yet. The identity of the department head, along with any connection they had to Evita's family, was a set of possibilities, not an already determined fact.

To see why everything must obey quantum superposition, let's imagine light that has just traveled from the Andromeda galaxy all the way to earth, a distance of two and a half million light-years. According to us, the light would take two and a half million years to reach us. We set up a telescope on a satellite orbiting earth that captures the light and then makes a random choice to reflect it either to a lab in Houston or a lab in Chicago. Remember, though, light is timeless, so its creation in Andromeda is one and the same with its absorption in either Houston or Chicago. But which is it—Houston or Chicago? If time had passed for the light between Andromeda and Earth, we could say, "We made a choice in between that affected things." But when the starting point and the endpoint are *one timeless event,* how does the light "know" which of these "realities"—reflection to Houston or reflection to Chicago— really happens, especially when mortals like us can make a free decision in between? Bohm's description of the implicate order is quite striking: "Indeed, in principle, this structure extends over the whole universe and over the whole past, with implications for the whole future."[120]

Without time to measure things by, we can't talk about things "changing in time," and it's hard to discuss free will in a meaningful way. Instead, we have to assume that all the possibilities already existed, and we are just choosing between them. The light must have a possibility of existing on either path, and these possibilities must "coexist." This is where we get the image of a timeless, branching tree of possibilities, made popular by the many-worlds interpretation of quantum mechanics. (See figure 18.) Bohm says, "So we are present everywhere and at all times, though only implicately. The same is true for every 'object.'"[121]

But this reasoning goes even further. In the example, the superposition of possibilities arises from the concept that light

Exploring the Foundations

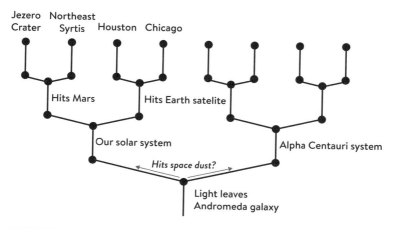

FIGURE 18. This diagram depicts a timeless, branching tree of possibilities of light leaving the Andromeda galaxy, bouncing off of a satellite orbiting Earth, and ending in a lab in either Houston or Chicago. At each branching, the light may get deflected by some obstacle in its path, such as space dust, and end up on either a left or a right branch. If it lands on Mars, it may hit a crater named Jezero or a region named Northeast Syrtis. We are interested in the case where it hits the satellite and a human's choice directs it. The light must have the possibility of existing on all of these paths, since they could all still occur when the (timeless) light left Andromeda.

is timeless. But anything that light touches—the detector in Houston, the detector in Chicago, your eyes, your skin—must also be in a superposition of possibilities. Even though the usual notion is to assume that the detector in Chicago or Houston must *cause* the light to choose a definite state, if the light is timeless, it's timeless. There's nothing more to it. It cannot be affected by a choice made by a human at a certain time. Without time, our traditional notions of choice, change, and free will must be adjusted. Bohm says, "Reality as a whole ... is also not to be regarded as conditioned.... It could not be consistently be so regarded, because the very term 'reality as a whole' implies that it contains all factors that could condition it and on which it could depend."[122]

Rather than light becoming "classical" when it meets a physicist, the physicist must become "quantum" when she meets the light. But doesn't the physicist have her own definiteness? Isn't she real and certain? Her state is definite *relative to her*, but she and the light are both indefinite *relative to me*.

Interactions, then, are relational.[123] Everything that interacts connects itself together, but that doesn't affect someone outside the room. If I am standing outside the room, although everything inside the room may be interacting and getting more and more tangled in mutual relationships, it's only when I walk in and observe any of it that it all becomes definite relative to me. We can no longer talk about the world from a hypothetical God's-eye view that describes everything at once without taking a specific perspective. That doesn't exist. We have to abandon an objective and definite overall perspective and limit ourselves to understanding the world only through an observer's eyes.

Any object that interacts with light, large or small, can be described as a superposition of possibilities. This is called a "macroscopic quantum superposition" state. This crazy idea is made famous by the story of Schrödinger's cat,[124] a fictional thought experiment where a cat in a sealed box may or may not have been killed by a random radioactive decay. According to the dilemma posed by quantum mechanics, until we look in the box, the cat cannot be said to be actually alive or actually dead. If we don't actually look in the box, what is the reality of the situation? The best that can be said is the cat is in a superposition of both options. Without an observation to pin it down for a specific observer, reality evolves into a branching tree of possibilities.

Retroactive Event Determination

There is one more important takeaway from our discussion about the timelessness of light. Our example also illustrates that when we interact with light, its actual reality from our perspective becomes determined right then and there. The entire two and a half million years of evolution of that light must conform to our experience of that light right now, and quantum mechanics tells us that the light's history is not determined until we see it. Therefore, its entire history retroactively falls into place.

I call this *retroactive event determination*. It avoids any paradoxes of what is usually called "retrocausality"—such as going back and time and killing our grandmother so that we cannot be born, in which case we could not have gone back to kill our grandmother in the first place—because we aren't changing the past. Instead, the past hadn't been decided yet, and our measurement simply made it what it was.[125] When applied to microscopic particles, this behavior is pretty well accepted, known as "Wheeler's delayed choice experiment." Yet due to the timelessness of light we just discussed, this appears to be true much more broadly—for all objects. While this position is still quite controversial within the scientific world, let's look at what it might mean in a real-life situation.

Think back on Evita's experience of getting into her graduate program. Because she had not interacted with the head of the physics department, my suggestion is that those circumstances were not yet established or definite from her perspective. When Evita picked up the phone and made the personal connection, she interacted with the department head. At that moment, specific properties of Evita's environment "fell into place" *from Evita's perspective*, including the history of the whole physics department. Imagine a history in which a different professor had been selected as department head that year, or a history in which this professor had had a terrible falling out with Evita's mother; these were both possible prior to Evita making the phone call. These histories, of course, would not have been very helpful for Evita's chances of getting into the program. It was a more useful set of circumstances that *did* end up falling into place, for reasons we will focus on in the next chapter.

Before we get there, let's ask ourselves about this phrase "falling into place." Is Evita really changing the history of the department head or her mother when she calls the school? That seems ludicrous, because (for those people) the events happened long ago and have been true ever since. But what about for Evita? She didn't have any information about who her mother's thesis advisor was, nor about who was assigned as department chair that year. If, from her perspective, there is no way of knowing what those histories were to begin with, can we really say that Evita's amazing coincidence was actually *changing* the past? No. If she could never possibly find out any information that contradicted the facts she now knows, then she has no right to claim that the past was different until she interacted with it in the present.

What's true is that the past was uncertain until *she determined what it had been.* And that's precisely what quantum mechanics tells us: things don't have established histories when we're not interacting with them. The labels *past, present,* and *future* become antiquated, and we instead talk about *determined* and *undetermined,* or *observed* and *unobserved.* What we call the past, unless we've observed it, is not actually established until needed, in order to line up with the specific experiences we are having. The world outside our walls is not yet shaped. *The end result makes the history fall into place.*

Gaming and Optimistic Synchronization

If you are a gamer, you might recognize the concept of "optimistic synchronization" in retroactive event determination and the relational view of the world. In massive multiplayer online video games, people from around the world compete and collaborate in virtual worlds. The game must be able to handle a wide range of computer speeds and internet speeds, including very long latency (lag time) as a result of the vast distances between gamers.

One strategy the software can use to accomplish this is not trying to track every detail of the virtual world for all players. In other words, just as described for relational quantum mechanics, it abandons the attempt to build a single world that is both objective and definite. Instead, it considers each player's point of view and calculates just what that player needs in order to make their game experience smooth. The gamer feels that they are in a vast virtual world, when in fact the details of their world only actually exist for the small local space around them.

When two players come into contact, the game must ensure that their experiences are consistent with each other. So optimistic synchronization is both relational (computed only locally for each individual user) and consistent (ensuring that all interactions between the virtual worlds of any players lead to shared experiences that make sense). Sometimes, if the player's selected motion hasn't been received yet due to poor connectivity, the software must guess what one player has already done in order to render it in "real time" for another player. This guess is "optimistic" because it is based on assuming that the player will continue what they are already doing, and it allows the virtual worlds of multiple players to synchronize adequately.

How does this work in practice? One example of a system using optimistic synchronization is the Time Warp Operating System (TWOS) described by Peter Reiher of the NASA Jet Propulsion Laboratory.[126] Generally speaking, an operating system in such a scenario will have to accomplish a lot with little input from users because the users are distributed far and wide, and of course they perform far fewer actions per second than a computer. When the game's operating system receives an action request from a user, it could be that the operating system's previous best-guess actions would have been out of sync with the current "real" action of a user. The operating system handles this by altering the previous actions and retroactively making them invalid. This is called a "rollback."

In a rollback, the operating system retroactively adjusts the unobserved past to remain consistent with the observed present. But if the past actions are already observed, then they can't be rolled back. Reiher reports, "TWOS simulations sometimes need to perform output to devices not directly under the control of TWOS [such as external disk drives]. Data written to devices not under TWOS control cannot be rolled back." So in the virtual world, data written outside the control of the operating system is permanent—or, we could say, observed.

In a real world described by retroactive event determination, there are many undetermined versions of the past actions—many branches of the tree—so the idea of rollback is a little different. The reason that the actions in the virtual world that have not been written externally can be rolled back is that nobody has seen those commands anyway. Were it not for the computer's log file, nobody would ever know that a change had been made. Analogously, in the quantum world, so long as you, the observer, have no way of knowing what any previous conditions of the world were, you will never know whether a rollback had occurred. In fact, due to the counterfactual indefiniteness mentioned in chapter 2, we cannot call it a rollback if we can't ever know what the previous state actually was. We can't call the previous state "actual" at all. In the quantum case, the rollback involves choosing (retroactively) between the various past histories available in the situation that are consistent with the current actions.

Exploring the Foundations

How does an operating system know when an event is observed? The main purpose of this method is to avoid conflicts between "real" events in the gaming world. To accomplish this,

> *TWOS must delay the actual I/O until the write request is certain to be correct. That certainty is obtained when the event performing the write is committed. Therefore, write requests are tagged with the virtual time of the event requesting them and are held until their commit point is reached.... In general, TWOS must delay performing any action it cannot undo until the commit point for that action is reached.*[127]

If we think about the "tag" as the time when an event took place, and the "commit point" as the time when the event retroactively falls into place and becomes known to have happened, we have a great practical model of a relational and consistent virtual world that mirrors the proposed process of retroactive event determination in the real world.

With this foundation, we can dive into the central ideas that I think give rise to measurable meaning and the emergence of synchronicity from science. If the science has felt boring or difficult so far, don't worry; it was intended as useful context, but it's not really required for what follows. Read on!

MEANINGFUL HISTORY SELECTION

Retroactive event determination tells us how the world *can* fall into place in many distinct ways as a response to the actions we take. But why does it choose a *specific* way of falling into place? Why does it sometimes seem like Murphy's Law goes into effect, causing me to hit traffic just when I most need to be on time? Why do things seem to always work out when I am "in the zone"? Sometimes life seems dead-set against me, and sometimes it seems like it is on my side. Is there a pattern at work here?

To help us look for this pattern, I am going to introduce more new ideas (both my ideas and those conceived by others), some of which are controversial to varying degrees. Some researchers examine what happens if we look at only the experiments in which a specific outcome occurs. This process of "post-selection" has been developed most fully in the version of quantum mechanics known as "two-state vector formalism." I suspect this concept leads (with some additions of my own) to the experience of meaningful coincidences.

Here's the fundamental idea: based on the timelessness of light, we find that all possible outcomes of a situation are immanently accessible to us in every moment. When we act, we choose the set of outcomes that could be actualized, and we also tag those possibilities that align with our anticipated qualitative experience. By doing so we are defining what is meaningful for us, and we shape the probabilities of the available outcomes. We cannot guarantee that the meaningful ones will happen, but we can tilt the odds.

As a reminder, in this theory there are some established ideas, some new ideas, and some speculation. For instance, it is well-established that all the properties discussed so far apply to microscopic particles. It is not yet accepted in the mainstream that these same principles apply to all objects regardless of size, though significant work has been done to suggest that this is true. Furthermore, the idea that qualitative experiences are a fundamental aspect of the cosmos is taken seriously by many philosophers, but the suggestion that this can be applied to physics is more speculative. Finally, the relationships between qualitative experiences, emotions, feelings, and thoughts are to some degree well-understood, but the idea that we naturally anticipate certain qualitative experiences and that this anticipation affects the physical world is a new proposal that I will argue for in this chapter.

The Tree of Possibilities

Here we come to the heart of the discussion. In the previous chapter I described how the universe is always evolving into many possibilities. Every object is interacting with other objects—like billiard balls rolling on a table or a person bumping into a friend at the grocery store—and when this happens there are many ways the interaction can end up.

Imagine that my friend Anne is on her way to a Broadway show from her apartment uptown. She has barely enough time to get there, and she knows that once the show starts they will not let her go to her seat. At each step in her journey she is interacting with objects around her (the escalator in the subway station,

Meaningful History Selection

the ticket vendor, the handbag and jacket she is carrying). Each interaction leads to a variety of possible outcomes branching off from a decision she makes: Should she take the stairs or the escalator? Which ticket line will move faster? Should she wear her jacket or wrap it around her waist?

She gets to the station, knowing that the train comes every twenty minutes, so that if she misses the next train she will certainly be late for the play. As she is buying her ticket, she hears the train pulling in. She is still waiting to get her credit card back, and she still needs to go through the turnstile and down an escalator. It is evident that there is no way for her to catch this train, right? What should she do?

One possibility is that she hastily retrieves her card and hustles to the turnstile, racing down the escalator against all hope, knowing that in all likelihood the train will pull away long before she gets to it. We'll call this branch R because she rushes. Another possibility is that Anne surrenders to her circumstances: she casually retrieves her credit card and walks leisurely to the turnstile, knowing the train will pull out at any second and she will have to miss the first half of the show. We'll call this branch $\bar{R}$ (pronounced "r-bar") because she does not rush. These two possibilities represent two branches of her tree of possibilities. (See figure 19.)

Each branch of the tree represents a particular configuration of all the things in Anne's world: the ticket vendor, the turnstile, the trains on the track, the other people in the station. Any two branches of the tree represent different ways that this collection of things could be arranged. In one branch, maybe the ticket vendor has trouble with the cash register and somebody is blocking the turnstile. In an alternate branch, maybe the ticket vendor moves quickly, and the turnstile is free and clear. Each time two objects interact, they lead to various unique ways the world could turn out, so each interaction is a branching of the tree.

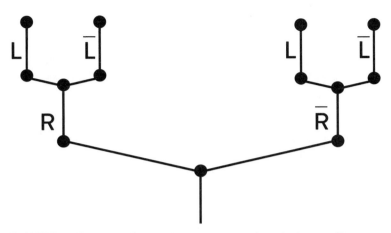

FIGURE 19. This tree of possibilities represents Anne's choices. She can either decide to run for the train (R) or not (R̄), and in either case she may or may not be late to the play. Each branch will, however, have different probabilities (or "weights") associated with it.

The overall point here is important, so I will restate it: a single branch does not represent a single object; rather, it represents *all* the objects within one possible "world," or one possible way that things could unfold. The result is an immensely complicated tree of possibilities that represents all the possible ways a set of objects can interact over time. However, not every branching point will have an effect on Anne's future. Fortunately, we can choose to draw only the properties we care about, which we'll call singular events. In the LORRAX process from chapter 3, this relates to the process of listening to the circumstances to discern which factors might be relevant to our goals.

Now let's put in some numbers. In figure 19, if Anne rushes *(R)*, it's reasonable to assume that there is at least a slightly higher chance she will arrive on time because the faster she moves, the greater the chances are that she will arrive earlier. On the branches leading off of *R* in figure 20, I've chosen the amplitudes 24 for being late *(L)* and 36 for not being late *($\bar{L}$)* to convey this. For illustrative purposes I've chosen numbers to associate with each branch that work properly with the actual physics calculations while also being easy to talk about. So right off the bat, Anne has a decent chance of not being late, because 36 is greater than 24.

But this is a very coarse description of Anne's trip to the theater. What about all the possible events that could happen in between her decision to rush and her arrival at the theater? Consider the three specific branch points in figure 21. We've expanded the branches to track a new property of Anne's journey. The first branch point is when she decides whether to rush for the train *(R)* or not to rush *($\bar{R}$)*. The last branch point involves her either being late *(L)* or not late *($\bar{L}$)* to the theater, as before.

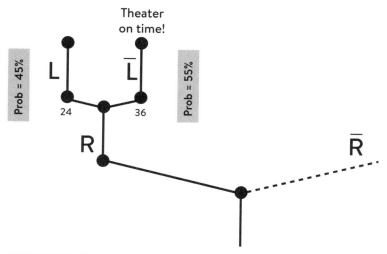

FIGURE 20. If Anne runs for the subway, there is a high weight (36) on the branch $\bar{L}$, where she is not late to the theater, while the weight on the branch where she is late is lower (24). Assuming nothing strange happens, we can assume that if Anne rushes for the train she is at least a little more likely to be on time ("not late") for the play.

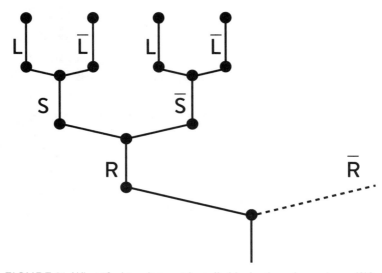

FIGURE 21. What if a bicyclist accidentally blocks the subway doors (S)? If that happens, it delays the train long enough to allow Anne to get on, assuming she ran to get there (i.e., if she is on the R branch). So on the R-S branch, she still may be late (L) or not late (L̄), but the chances are higher that she won't be late.

The events S and $\bar{S}$ represent an *intermediate* event. For instance, let's focus on the subway train that is already sitting in the station. Anne's own logic tells her there is no way to catch that train. But hidden in her logic is an assumption about the way things will go. The train has not yet left, so she has not actually missed it yet.

If we think creatively, we could imagine any number of unusual circumstances arising that might delay the train from leaving the station long enough for her to get on board. Let's say a bicyclist happens to be getting on a very crowded subway car and can't quite get her bike inside the doors. This keeps the engineer from closing the doors for an extra thirty seconds. We'll call this event S for "subway" or maybe for "synchronicity." If S happens, and if Anne has rushed *(R)* to get to the train, then amazingly she will catch the train and make it to the play on time $(\bar{L})!$ The opposite experience, in which nothing unusual happens, is labeled $(\bar{S})$ and most likely leads to her being late for the show, just by default. We're now tracking a new property S of Anne's journey. Figures 19 and 21 describe the same physical situation, with the second image simply having more detail in it.

Now, let's think about the bicyclist getting caught in the door. It seems obvious that Anne's decision to run from the ticket vendor to the train platform certainly can't influence the bicyclist to show up when she did, for the bicyclist is already at the subway platform and Anne is not. Figure 22 shows that if Anne rushes, there is, let's

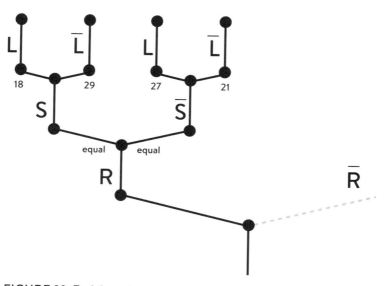

FIGURE 22. Each branch off of R has equal weight, and Anne's decision to run shouldn't have any impact on whether the bicyclist delays the train. If the bicyclist does delay the train (S), the branch with the highest weight (29) is the one on which Anne is not late $(\bar{L})$ to the play. If the train is not delayed $(\bar{S})$, the branch with the highest weight (27) is the one on which she is late (L), because in this case she misses the train altogether.

say, an equal likelihood of the bicyclist blocking the subway doors or not. Anne's choice cannot causally affect these likelihoods. We would conclude that if the bike gets stuck it's just a fluke, whether or not it helps Anne get to the train in time.

But the same is not true for the next layer of branches, from the S level to the L level. If Anne rushes *(R)* and the bike gets stuck *(S)*, then Anne catches the train, so she will very likely get to the play on time. The amplitude 29 on the branch R-S-$\bar{L}$ (compared to 18 on the branch R-S-L) shows this. You can also see that if the bicyclist *doesn't* get stuck, and therefore Anne *misses* the subway, the opposite is true. The amplitude on the other branch *(R-$\bar{S}$-L)* in which the bicyclist doesn't get stuck and Anne is late to the theater is quite high (27). This is just as we would logically expect.

So let's gather together what we know so far. From figure 20 we know that rushing and arriving to the play on time are somewhat related. From figure 22 we know that a bicyclist delaying the departure of the train and Anne arriving to the play on time are somewhat related. But we also see from this figure that Anne rushing is not related to whether the bicyclist delays the departure of the train.

We can now pose the crucial question: "Does the helpful coincidence of a bicyclist getting stuck in the subway door become more likely if Anne makes an effort to catch the train?" In other words, if Anne rushes for the train out of her authentic yearning to make it to the play on time, is the most useful history more likely to fall into place? The proposal of my work is that this is indeed the case, and we will see in a moment how this could be.

Before pressing forward, let's keep in mind that branch S is not only referring to the one circumstance we talked about, where a bicyclist blocks the train doors. Branch S could represent any type of event that is more likely to lead to Anne being on time to the theater. For instance, S could represent a situation where the lead actor in the play arrives late at the theater, so the play itself is delayed, therefore allowing Anne to get inside the theater before the doors close. ($\overline{S}$ would then be the case where this doesn't happen.) So long as S is a circumstance that is biased toward Anne's anticipated qualitative experience of arriving on time for the show, we would call it a synchronicity.

> ## More Valuable Than Gold
>
> Barnett Rosenberg was a chemist investigating the effect of electric current on the growth of bacteria. He accidentally discovered that his bacteria kept growing but couldn't replicate. Through careful analysis he found that the set of platinum electrodes he was using in the petri dish were corroding into a compound called cisplatin. One can make electrodes out of gold or a host of other materials, but Rosenberg's choice of platinum and his synchronistic discovery led to the use of cisplatin as an anticancer drug.[128]

If you haven't really followed the numbers in the diagrams, don't worry one bit! Some people relate to the numbers, others relate to the stories, and either way of thinking can bring you a good sense of how synchronicity seems to work. The idea so far is that the numbers capture the qualitative relationships we expect between Anne's possible experiences. Up until figure 22, no "synchronistic" effects have happened. We have simply written down "this leads to that." Now we will go a step further and examine how the bicyclist inadvertently delaying the train can be seen as the cosmos responding to Anne's choices.

Selecting the Anticipated Qualitative Experience

In chapter 2, I suggested that properties and qualitative experiences can be related. For example, the property $\bar{L}$, which is a symbol representing a branch on the tree where Anne is not late for the play, is the same thing as the experience Anne has when the property $\bar{L}$ happens. In other words, $\bar{L}$ can be looked at as either a physical property of Anne and her surroundings or as a type of experience she has.

How do we define that "experience"? Well, being at the play evokes a certain set of thoughts, feelings, and emotions related to the experience. There are the vibrant visuals that evoke an emotional response through her eyes; there is the dramatic soundscape that stimulates her ears; there is the salty taste of popcorn on her tongue; there is the feeling of intimacy from being at a special event with a friend; and there might be a feeling of awe that she is doing something quite special. All of these experiences—and the corresponding emotions, feelings, and thoughts—exist only on the $\bar{L}$ branch. If she arrives late to the play *(L)*, these things do not happen. So in looking at the branch $R\text{-}S\text{-}\bar{L}$ on the tree in figure 22, we think of it as "Anne decides to rush *(R),*" "the train is delayed *(S),*" and "Anne is on time to the play *($\bar{L}$).*" Yet all three of these describe *qualitative* events that could actually happen in many different ways and still be *qualitatively* the same.

How do we select specific branches of the tree to actualize? This is related to the measurement problem in quantum mechanics, which we talked about in the previous chapter (and will discuss in appendix B). It's a substantial unresolved mystery in physics, but we won't dive further into the details here. What I mentioned in the last chapter is that our "question" defines the set of possible answers that we can get from the world, and then the system supposedly tells us which of those possible answers is true.

If the theory of meaningful history selection is right, the plot goes one level deeper. Anne's action really has two elements in it: the physical action itself plus the anticipated qualitative experience that motivates the action. First of all, her physical action moves her along the branches of the tree by virtue of her choice to run through the train station. This action expands the tree into the R and $\bar{R}$ branches, and it puts her on the R branch. Although the other branch, $\bar{R}$, still exists in the bigger picture of the tree, from her view she sees only a single branch R of the tree. According to relationality, we have to define things *from her perspective*, and she has to experience a single branch. By choosing to run across the subway station, she trims off the branches on which she doesn't run. By making choices, she navigates her way physically through the world and metaphorically along the branches of the tree.

But what about the "anticipated qualitative experience" part? The theory is that we, as conscious beings, are always anticipating what qualitative experiences (or qualia) we may have as a result of our actions. Our ability to *feel* a potential experience while it isn't actually happening yet—to anticipate a future qualitative experience—affects the tree of possibilities and shapes our world.

This happens in two ways. First of all, it groups the branches in the tree according to whether the anticipated qualitative experience happens or not. I suspect this is related to what physicists call "choosing a basis." Her anticipated qualitative experience gives her the branches "late" and "not late" in the first place, instead of some other groups of branches like "love" and "not love," which might describe whether on this particular night she meets the love of her life or not.

If she is anticipating a future experience like salty popcorn on her tongue, "salty" is most likely to happen if she is not late to the theater, so this anticipated qualitative experience matches with $\bar{L}$. If instead she is anticipating the qualitative experience of "falling in love," the same underlying branches exist because the possible outcomes haven't changed, but they are grouped into L and $\bar{L}$ in a different way. $\bar{L}$ won't correspond to being on time to the theater anymore. If her anticipated qualitative experience is to meet true love, there's no reason to think this will be more likely if she is on time to the theater.

Here we need to understand a little about how physicists would compare two properties, or two branches. A mathematical procedure called an "inner product" measures the overlap between two states. If you and I show up at a party and our clothing happens to be color-coordinated, we could imagine calculating an "inner product" by comparing our clothing piece by piece. This would tell us how similar our outfits are. We might find that we are 75 percent alike, since most of our clothing matches except for our hats and socks.

Because the tree of possibilities is timeless like the light particles we talked about, meaningful history selection proposes that we can calculate the overlap between the *present* branches of the tree and the *future* branches of the tree. In other words, we can timelessly compare our actions now to their possible future consequences. It is as if whatever we choose ripples forward on the branches of the tree until it hits the leaves and then tells us how much each leaf aligns with our chosen action. These become the apples.

Back to Anne's subway trip to the theater: the event "being on time to the theater" is composed of simple, concrete experiences—like salty popcorn on her tongue—that she might have at the theater. If that's so, then any outcome in which Anne has those experiences will count as "the theater experience." For instance, even though she is late, she might still be allowed into the theater for standing room only, in which case she still has many of those experiences. Alternately, she may miss the play only to find that a movie theater next door is showing a movie she wants to see, so she still gets the salty taste of popcorn and the beautiful sights and sounds. These still count as meaningful outcomes, in alignment with her anticipated qualitative experiences.

When we understand the tree as symbolic, rather than a literal representation of the physical world, we might not be so disturbed by the idea that it represents many versions of reality. They are not *actual* alternate realities, only possible ones represented symbolically as a tree. Once the leaves of the tree tell us how much overlap there is between a particular branch and the action Anne has taken, we can group these possibilities into meaningful collections. For instance, all of the outcomes in which Anne ends up eating salty popcorn could be grouped into a single branch, maybe labeled *SP*. There will also be a nearby branch in which *SP* doesn't occur.

Here, finally, is the key step: a meaningful history is selected. When Anne takes an action like "rushing for the train," she does so because she is seeking to have certain experiences. She wants to be on time because she wants all the sights, sounds, tastes, smells, and feelings that will happen at the theater. Her action is both literal and symbolic. It is literal because when she rushes she actually moves her body and gets closer to the train. But it is also symbolic, in that her action aligns with certain qualitative outcomes on the tree. When she rushes because of experiences she wants to have, she is motivated by the statement "I want to be on a branch on which those experiences happen!"

On the tree of possibilities, the future is not some far-away concept. The tree is timeless, so every action Anne takes is acting on the entire set of future possibilities, not just the here and now. The possible futures are immanently accessible to her. Anne's choice to run acts upon the entire tree, and since she is most likely to taste salty popcorn if she is on time to the theater, there is a strong bias toward that outcome. Her symbolic action, rushing for the train *(R)*, will be symbolically aligned with the experience she wants to have, being on time to the theater *($\bar{L}$)*. This is where her choice to rush selects out a certain qualitative end result. In figure 23, the numbers 21 and 43 at the top leaves of the tree indicate that her "rushing" action is biased toward the "not being late" branches *($\bar{L}$)*, because 43 is bigger than 21. We multiply the number from her symbolic choice to rush (43) by the amplitude that already existed on that branch (29) to obtain a fairly large number, represented by a large apple on the branch. Not only was this branch fairly likely to begin with; it also received a bias from Anne's choice to run. It matches her anticipated qualitative experience.

The effect of the apples is to shift the weight of each branch. Some of the apples are bigger than the others, and this weight corresponds to the probability that the branch will happen. Still, nothing has been decided. Even the branches with lots of apples may be surrounded by other branches that have no apples, and these dilute the chances of obtaining the anticipated outcome. We can calculate the weight of a branch due to all the apples on it, and we can repeat this calculation for the branch holding up *that* branch, and then for the branch holding up *that* branch, and so on. At the trunk, the bottom branches carry the weight of all the branches above them, and this weight corresponds to the probability that the branch will happen. (See figure 24.) The branches of the tree have become weighted in favor of *S*, partially influenced by their alignment with Anne's choice of action. She has shaped her world by shifting the probabilities of the possible outcomes available to her.

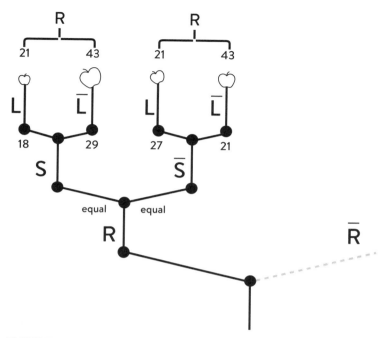

FIGURE 23. Anne's decision to run for the subway is motivated by the anticipated qualitative experience she wants to have: the sights, sounds, smells, and feelings she will experience at the theater. This acts upon the top branches of the tree, applying a bias toward the branches on which she is not late ($\bar{L}$) (43) over those on which she is late (L) (21).

As a side note, the concept of anticipated qualitative experience may relate to what some people think of as "intention." I feel that anticipated qualitative experience is a more accurate term. For one thing, it allows for the possibility that we may be anticipating a qualitative experience without consciously realizing it, whereas intention seems to imply a conscious motive. I suspect that synchronicities frequently arise in connection with anticipated qualitative experiences that we are unaware of, in which case we may not recognize the synchronistic nature of the occurrences. If we focus only on our conscious intention, we would have to conclude that our intention had no impact on the final outcome. But if we look deeper and recognize that we had a strong emotional pull toward or aversion to a circumstance, we might see that we were indeed anticipating the qualitative experience that actually unfolded.

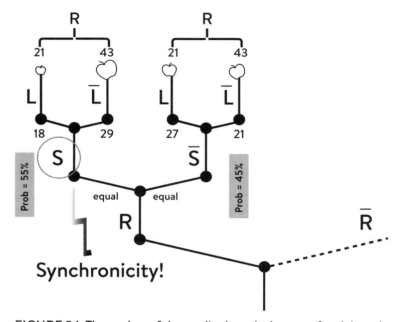

FIGURE 24. The product of the amplitude at the bottom of each branch (e.g., 29) and the amplitude at the top (e.g., 43) is the total weight for each branch. Big weights get big apples, and small weights get small apples. If we average the weights of all apples on a region of the tree, we find the weight on the lower branches. In this example, the apples on the left side weigh more than those on the right, and the S branch has a higher probability of occurring, even though originally the probability of S and S̄ were each equal. Therefore, as a result of Anne's choice to run, there is a greater likelihood that the bicyclist will block the door long enough for her to get on the train. This is a synchronicity.

Also, the phrase "anticipated qualitative experience" describes exactly what is happening: if we accept the idea that experiences are the building blocks of reality, then we have the ability to influence which events occur based on what type of qualitative experience we are anticipating or feeling in our body and mind. The word "intention" doesn't explain itself in the same way.

The Physics of Synchronicity

Figure 24 is where synchronicity happens. The branch S represents any situation in which the train is delayed long enough for Anne to get on. The branches R-S and R-$\bar{S}$ have equal weight, which means that if Anne rushes (R) she has an equal probability of ending up at S or $\bar{S}$. In other words, Anne should expect a 50 percent chance of experiencing a beneficial train delay that allows her to catch it. But after she has rushed for the train and apples have grown on the tree that reflect her action, that number has been raised to 55 percent. This may seem like a small increase, yet it is the main point of this book. The S branch becomes more likely because it itself will be more likely to lead to the selected symbolic outcome, "being on time to the theater." (Or, more accurately, the S branch will be more likely to lead to the qualitative experiences Anne will have—popcorn, sights and sounds, friends—if she is on time to the theater.)

A circumstance like a bicyclist blocking the train doors from closing, represented by S, is a meaningful coincidence, or synchronicity. There may be many situations that could delay the train and allow her to get on, and we could call them all S. Drawing the tree with any of the different possible S's still ends in a similar result: the alternate branch S becomes more likely because it, too, leads to her being on time to the theater. *We come to the conclusion that any meaningful coincidence, or synchronicity, becomes more likely when a person acts in alignment with a set of experiences she or he seeks to have.* This is not the only possible definition of synchronicity, but is the one we will stick with for the purpose of this book.

The synchronistic event (a bicyclist delaying the train) can be termed "meaningful." It is not related to Anne's action directly through cause and effect, but it *is* related to the future experience that she anticipates. This might seem like Anne is causing the bicyclist to get stuck in the doors through some psychic power, but this is not the case. Anne's choice cannot *cause* the person with the bicycle to have a particular experience. However, because Anne's world is defined relative to Anne's point of view, the bicyclist's particular location at that moment can fall into place in any number of ways. Our tendency to rely on supernatural powers to explain synchronicity comes from our ingrained assumption that the world outside of us exists in a definite form regardless of our perspective on it. From that worldview, if the train is delayed, it is either pure luck or Anne has the supernatural power to influence the bicyclist.

The alternate worldview that I suspect physics leads us to is that the world is relational. It is defined always from a particular point of view—yours—and the events that happen within that point of view just need to remain consistent with all the other points of view when we compare them.

So Anne's choices shift the probability of each possible history of the bicyclist. Still, she is not guaranteed a meaningful outcome. Rather, the synchronicity becomes more likely to occur, and if it does, she has the chance to respond effectively to it or not. We can see in figure 25 that if the synchronicity occurs, then Anne finds herself able to get on the subway. This branch has a 72 percent likelihood of leading to her being on time to the theater, based upon squaring the relative amplitudes (29 and 18) of the branches. This is a large increase from the probability of 55 percent in figure 20, which was based just on normal causality.

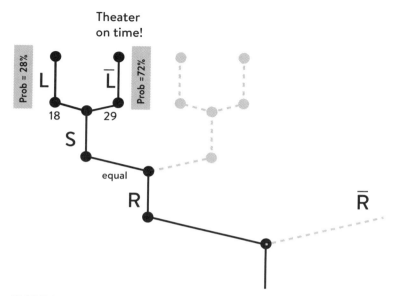

FIGURE 25. Once the subway is delayed and Anne gets on, the situation is now in her favor. On this branch there is a much higher likelihood that the $\bar{L}$ branch occurs and she is not late. The original probability for the $\bar{L}$ branch (in figure 20) was 55 percent, but now it has become 72 percent.

Even though I have described a situation in which the synchronicity was a positive, helpful circumstance, this doesn't have to be the case. We could change the example to have a negative consequence. Let's say Anne leaves the house with plenty of time for the show. Because she thinks she has plenty of time, she dawdles at the curb by checking her mail on her phone. Then she realizes she needs a warmer jacket, so she walks back into the house. Now she's a little rushed, and meaningful history selection is working against her. Her lack of urgency biases the situation toward the L branches because she is focused on other tasks and not actively anticipating the experience of being at the show. It might even be that she is acting upon a feeling like "I am nervous to be out with these people," or "I'm not sure my friends will really care if I am there," or "I haven't really earned a fun night out," which is making her more likely to hesitate. Whatever the inner story, something is causing her to minimize the importance of getting to the play on time, and the resulting bias toward the L branches switches the numbers around. There is now a higher chance that any event will occur that leads to the L branches. For instance, let's say the subway station she goes to is unexpectedly under construction, and in the time it takes to walk to the next station, she misses the train she needed to catch. This is a synchronicity that makes her underlying anticipated qualitative experience more likely to occur, even though it's not the experience she thought she wanted.

Meaningful history selection, as I see it, is how the cosmos responds to us as cocreators of life. It is neither friendly nor unfriendly, neither good nor bad. It reflects back to us the experiences that our actions are seeking. Sometimes we are conscious of the anticipated experiences guiding our actions, and other times we are not. This doesn't change how responsive the cosmos is, but it is a big factor in whether we feel like the universe is friendly or not. If we are conscious of what we are trying to achieve with our actions, then the result of meaningful history selection might more often feel good to us. We might more often feel like the universe is on our side. Contrarily, if we are unconsciously anticipating a different type of experience than we realize, the result of meaningful history selection might more often seem totally frustrating.

However, we should exercise caution when interpreting the responsiveness of the cosmos. Frustrating or painful experiences do not necessarily mean we have an underlying subconscious motive that is sabotaging us. Life is full of challenging circumstances, and we are not the cause of all of them. What I recommend is that we use an understanding of this process to see how we might get into flow, align with our circumstances whatever they may be, and help evolve them for the better starting from where we are.

To summarize the main scientific idea of the book, the process of meaningful history selection gives rise to the experience of synchronicity. When we interact with the world around us, we interact with the entire tree of possibilities, including all the future possible states of the world. Some of the future states contain experiences that align with our actions and are "selected," meaning that they have some weight—an apple—applied to them. But this does not mean we know the future for certain, because these outcomes are surrounded by other branches with no apples. The meaningful outcomes get diluted, but if the overall weight increases on one branch more than the others, that branch becomes more likely. We label these intermediate experiences synchronicities because they lead toward branches with apples. They lead toward outcomes that are aligned with our own choices and intentions. These experiences shape our world because they dramatically affect the future outcomes. They feel meaningful and they help us along whatever path we've chosen.

Jung's Acausal Connecting Principle

As a very minor illustration of this connection, I was recently due for an event eight minutes away by a rural road, and I gave myself nine minutes to get there, which is usually fine. Along the way, a police car pulled in front of me, which ensured that I would be sticking to the speed limit all the way there. Next we got stuck behind a city bus, and with a police car in front of me, I was not about to try to pass. I just had to sit tight while we crept along the country road. These events are synchronicities because, even though I technically had enough time on a normal day, the cop and the bus led me to have the experience—being late—that aligned with my lack of urgency. In this case, the events are meaningful in a Murphy's Law sort of way. I drive that route frequently and haven't had a similar experience with a bus or police officer before or since. I suggest that on this particular day, having taken for granted that I would have enough time to get to this important appointment, my actions had an objective meaning that aligned with the unexpected obstacles I encountered.

This is not a strong example of synchronicity, and I don't take it too seriously, but it illustrates synchronicity's neutral quality. This is often overlooked when people talk about synchronicity. Defined in this way, synchronicities are not good or bad events. Rather, they are events that align symbolically with the choices we make. They form a connection between our actions in the present and our intended outcomes in the future. Jung says synchronicity is an "acausal connecting principle." By "acausal," he means that the events are juxtaposed in a meaningful way but there is no way for one event to lead to the other. He seemed to intuitively grasp the timeless nature of light, as we discussed in chapter 7: "Since … under certain conditions space and time can be reduced almost to zero, causality disappears along with them. For this reason synchronistic phenomena cannot in principle be associated with any conceptions of causality."[129]

Because the process is not causal, we ourselves are not "making synchronicity happen." All we are doing is choosing to act in a way that biases the tree of possibilities in the direction we want to go. We can't use synchronicity to achieve a specific outcome we desire. Rather, a given synchronistic experience can happen in any number of different ways. Fate follows free will. Meaningful coincidences happen in response to what we choose. Anne doesn't try to get the train to be delayed; she simply tries everything she knows to do to get to the theater on time. A synchronicity is

Meaningful History Selection

> ### I'm Sorry, Who Are You Again?
>
> "One day when I was visiting Austin, Texas, I totally forgot about a lunch meeting I had scheduled with someone who lives there. I went to a restaurant to get some lunch by myself, and a few minutes after I sat down a man walked up to my table and said, 'Hey, sorry I'm late.' It took me a second to recognize him, but then I realized he was the person I was supposed to meet! Somehow I had ended up deciding to eat at the same place and the same time we had agreed to meet." (Story contributed by Mikey Siegel)

always an unexpected event that emerges out of the tree of possibilities—not an event we planned on, but one that serves our purpose. We do not cause it, but its likelihood increases acausally as a result of our choices. Synchronicities usually stand out as events that seem rather unlikely compared to the usual random background noise of life.

Zukav discusses this within the context of the soul's journey in a way that reflects the branching tree of possibilities: "Whereas [a specific] possibility might have existed with only a small probability at the time of the soul's incarnation, a door that would open only under certain circumstances—if this happened and then if that happened—it may become so that the soul does indeed find its way to that path."[130] The set of choices we make influences our path through the tree, not simply by controlling the outcomes of life but through the meaning of those choices.

Let us again consider one of the questions at the opening of chapter 2: do our actions have a purpose? Each possible synchronicity in a certain situation may involve different configurations of the world, but these will all lead to a similar meaning. We may end up being a doctor, soldier, teacher, artist, athlete, or businesswoman, and the branching point differentiating these life paths may be only one momentary choice that in retrospect determined the whole direction of our life. Yet these different paths can each be meaningful in a similar way, bringing us in contact with the lessons we need to learn and the satisfaction we are seeking. Living in a world of synchronicity means being less focused on specific physical circumstances and more focused on how the circumstances align with our core values and wishes.

Synchronicity is a neutral process by which we get back more of what we put out. The world brings us events that reflect our actions, and because our actions generally follow from our belief systems, this theory implies that our belief systems end up being a blueprint for the sorts of experience we have. Meaningful history selection is simply a mechanism for this to happen. It appears to be the mechanism that Jung referred to as an acausal connecting principle.[131] It plays a similar role to cause and effect but is based on the meaning of events—the symbolic language of the tree of possibilities—rather than the actual interactions between things.

The Library of Heaven

Another way to think of the tree of possibilities is like a library. Imagine yourself walking through a grand library, with vaulted ceilings and towering rows of books stretching out in all directions. Each book you see contains its own story, telling about adventures or sorrows, mishaps or celebrations. The sections of the library are organized by genre; there are fiction and nonfiction, romances, scientific biographies, cookbooks.

You are standing in an aisle between two of these tall shelves, surrounded by books in a particular genre, looking at a selection of books related by theme or author. From where you stand there are a hundred or so books you could choose.

The tree of possibilities is like this. Certain experiences are possible from your present vantage point in the tree. Other experiences are not possible from where you stand, in the same way that you won't be able to find a book on auto racing when you're in the political philosophy section. The tree of possibilities is like a library, where each path you walk is a book on the shelf. The books represent all the possible outcomes of every possible scenario. They are the branches of the tree, representing all the possibilities for the entire universe.

This is, in my mind, the scientists' definition of heaven. I use this word in the sense of "the totality of possibilities," rather than in the sense of some kind of transcendental perfection. We can think of the tree of possibilities as a library of heaven. It is not an Amazon website where we can search for and purchase any book we want without leaving our seat. In the library of heaven, we must physically move through the shelves in order to find the books we seek. Each event in our lives is a splitting of the tree, so we move through the library whenever things happen to us. Our response to each event in our lives influences which area of the library we move to.

In the library of heaven, all possibilities exist, but clearly not every possibility is accessible from any given vantage point. When someone says "everything is possible," they seem to be drawing on the notion that quantum mechanics is a theory of "infinite possibilities." But the possibilities of quantum mechanics are not infinite, and they are not equally accessible from everywhere. We are constrained by the section of the library we are currently in.

I think the true meaning of that phrase is that every *type* of symbolic experience is possible. Through meaningful history selection, it may always be possible to envision a certain type of experience, say being a doctor, and then to take steps to make that type of experience real for ourselves. Someone who is eighty years old and has always dreamed of being a doctor may not realistically be able to go to medical school and become a professional doctor. Yet they may be able to satisfy the same purpose for themselves by volunteering at a local hospital or becoming a caretaker for somebody they know. It is not always possible to create a specific desired situation from within our current circumstances. Certain opportunities in the past have passed us by, and they don't return again in the same form. Yet other versions of the experience may still be attainable with the proper actions and enough patience, even if a specific opportunity has passed. To achieve a particular type of experience, we have to move skillfully through the library. Not everything is possible, but with meaningful history selection it does appear that much more is possible than we might think.

YOU ARE A SPARK!

The role of a change maker is to take small risks that catalyze the energy in a set of circumstances and see which synchronicities emerge. Sparking meaningful situations is a direct path to living in flow. The modern world is full of situations that can be ignited into moments of creativity and connection. Any mundane office meeting has moments where one person showing up authentically can shift the level of human connection and maybe even lead to innovative ideas. Bringing up something that is authentically important to us ("Can we practice passing around a stick to make sure everyone gets a chance to talk?") or offering an idea that might seem unusual ("Do you want to move the meeting to the outside patio?") are great places to start. Family gatherings, subways, parties, public spaces, concerts … we experience no lack of opportunities to light a spark if we listen for them. By looking to our own inner guidance, we can develop the discernment to tell which opportunities serve us and the others in our lives.

The spark we initiate is an action that leads toward the type of symbolic outcome we wish to achieve (e.g., feeling comfortable being ourselves at work), and it lays out the apples in a meaningful way (i.e., biased toward events in which this happens). Lighting a spark often feels a little risky, which indicates that we are putting energy into a situation to change its trajectory. We never know precisely what will emerge from our spark, but when we recognize potentially positive situations and step boldly into the unknown, we can trust that some kind of useful circumstance will unfold. We can't control the fire, but we can influence the direction of the burn. I believe our job is to be a catalyst for the experiences we wish to see more of. Our job is to show up.

My life as a musician is a steady source of reminders of the power—or necessity—of lighting a spark in order to blaze a trail. When I visited my sister and her baby in Minneapolis, as I described in chapter 4, I posted a message on social media to find musical connections in the area. A person named Mari replied, a friend of a friend whom I had never met, and invited me to go to a music party the next evening.

Getting the Right Final Outcome

When I was attending my graduate program in physics at San Francisco State University, I commuted two hours by bus to get to school. One day I needed the two-hour bus ride to catch up on schoolwork, but I couldn't find my keys to drive to the bus stop! I had left them in Dana's car, and I knew she might be home soon, but if I waited too long for her I would miss the bus. My friend Tom, a bus connoisseur, texted me the location of a hidden bus stop I could go to if I could only beat the bus there. Dana got home just in time for me to get my keys from her. I sped off and arrived at the platform before the bus did. One problem solved! But when the bus arrived it was completely packed, with barely any standing room, so it seemed there was no way I could study. I edged my way to the back to find someone taking up an extra seat with their backpack. They cleared it off and I sat down to work. I arrived at school on time and caught up.

It was a classy local jazz event, hosted by a patron of the arts. As I followed Mari into the parlor, I took in a few notable synchronicity clues. She had brought us in through a special door that put us right in the front row. I was in a seat about three feet from a baby grand piano, and next to the piano was an electric keyboard.

I'm pretty shy about sharing my music in public, so to stretch myself out of my comfort zone I try to say "yes" when life offers unexpected opportunities. These circumstances put me on the alert.

The first song they played had chords I could easily follow by ear, and the electric keyboard was empty. I checked in with myself—to make sure I was acting from a sense of play rather than from ego—and then I stepped up on stage and jumped into the song. I took a solo after the other pianist and got into the groove with the band. Afterward I smiled a little sheepishly and returned to my seat.

Mari was slightly flummoxed. She leaned over and said, "Nice work on jumping in and following the flow! But next time please check with the host first." I was a stranger, and I understood why the host might be nervous about my unannounced participation. I was beginning to feel badly, fearing that I hadn't really lit a spark and had instead made a spectacle of myself.

But on the next song, the lead singer looked directly at me in the audience and said "Aren't you going to sing this one with us?" I was so taken by surprise that I didn't have time to question. It was either do it or don't. So I jumped up and joined in. Clearly my bold step had catalyzed something unexpected.

After the show, the host invited me to play the piano while folks were milling in the lobby. As I sang "My Romance," the lead singer walked back in from the lobby and said, "I want to play that one with you." From this little spark, an hour-long jam session unfolded that included the whole band. This time, I was part of it all.

What would have happened if I hadn't lit a spark? My night probably would have been pleasant; I would have made a few friends and appreciated the performance I saw. My life would be basically the same afterward as it was before. But because of that spark, I bonded with a new community of friends and professionals. I was part of a fulfilling musical experience. I have a priceless memory from the trip. And Mari has become my new vocal coach. My life was significantly changed by that night, and my trust in my inner knowing is stronger from the experience.

We are all creative, but we don't just create *things*, like music, art, science, or commercial products. We can use synchronicity to create *situations*. There is no more powerful way to change the world than to create situations. Educators know this well, because student learning cannot be forced; it can only be invited and sparked. What rift in the world could you imagine healing if only you could create the right situation? What problem could you solve if only you could get the right people in the room together? What part of your life could you have more success in if you could find yourself in the right conditions?

We can return now to the opening questions: What is the meaning of life? Do our actions have a purpose? Is the universe friendly? In my view, the purpose of life is to create meaningful situations that make the responsive cosmos a friendlier place.

Everyone Wins in Flow Consciousness

The LORRAX process can help us spark more unique situations and find our flow. We can *listen* to the events of our lives, *open* our minds to what adventures or creative possibilities they may carry, *reflect* on them long enough to learn something new, *release* our preconceptions about what we should or shouldn't do, *act* boldly, and *(X)* never stop the process of learning from life. Living in flow is a dynamic process in which we don't know exactly what will be required next. It means staying in relationship with life under any circumstances and surrendering our belief that we know for sure what is true. In Joseph Jaworski's words, it means "listening to what is wanting to emerge in the world, and then having the courage to do what is required."[132] It does not mean adopting a wishy-washy relativism; living in flow is a commitment to reevaluating our understanding at each moment, seeing each situation with fresh eyes, and hearing each conversation with fresh ears. This is how science works when it is at its best, and it is also the defining feature of a spiritual approach to life at its best.

Just as water avoids the extreme peaks and flows instead in the middle of a valley, flow lives on a middle path. On this path, we honor our outer need for material abundance for what it can teach us and what delights it can bring. We also honor our inner need for meaningful experiences, for the richness and color they bring to our lives. Without denying our own needs, we can find things to do that bring more love and care into the world. By living from the heart with boldness, gratitude, and self-knowing, flow helps us find a deeper sense of dynamic safety in the world even when the outer circumstances feel uncertain. The dynamic safety comes from following our inner compass to decide which events are meaningful and which flow we choose to enter.

Stepping fully into our lives in this manner, each of us can be a part of a shift to flow consciousness that can support a worldwide renaissance of creativity. This shift has already started, as more and more people build symbolic momentum in the direction of a world that works for and honors everyone. In flow consciousness our choices actively shape the types of experiences that make up our world, but we let go of worry about controlling the circumstances themselves.

Jaworski sees it this way: "If individuals and organizations operate from the generative orientation, from possibility rather than resignation, we can *create* the future into which we are living, as opposed to merely reacting to it when we get there."[133] As climate change progresses and the demands of survival become more pressing, my guess is that we will find ourselves naturally relying on each other more. As we do, I hope we will outgrow the habits of vying for attention or resources. Flow allows us to participate in an abundant universe in which we know—we feel deep inside—that there is enough for us. When we rely on flow for our success, we are not attached to a specific way of doing things. Because there are many possible branches in which we are successful, we no longer see other people as our competitors. We see foremost their humanity, and secondly the thrill of the competitive spark that is driving us all to become our best selves. The incentive to be underhanded is decreased because we can achieve our goals through flow rather than force. Constraints, conflicts, and complications become positive driving forces for innovation.

Living in flow can be seen as a practical path or as a spiritual path, whichever suits our individual temperament. In talking about businesses that have achieved a measure of self-organization, Laloux says such a shift often is accompanied by "a profound sense that … we are all connected and part of one big whole. After many successive steps of disidentification, as we learn to be fully independent and true to ourselves, it dawns on us that, paradoxically, we are profoundly part of everything." People naturally seek "to find wholeness with themselves and with others."[134]

We can get into flow and make a difference in the world by developing ourselves and overcoming our own demons. The world we experience out there is a reflection of the world inside ourselves, and when we are willing to show up authentically, we naturally dissolve problems. Just as with my home construction project gone wrong from chapter 2, my hope is that when we allow ourselves to live in flow by genuinely feeling our experiences and finding a dynamic balance between pushing and letting go, we experience an increased sense of well-being—not only our own but that of the larger collective as well. In today's either–or America, some people believe in the power of the individual, and others believe in the power of the collective. But in flow, the success of the whole is aligned with the success of the individual.

This, it seems to me, is the great mystery that transcends the old paradigm and leads us to a new era. Is it possible to find a way of being that benefits others while also benefiting myself? Is it possible to engage in a path of personal and professional fulfillment that simultaneously addresses communal problems such as climate change, social justice, or any other issue you care about? I suspect that when we live in flow, the answer is "yes."

The Synchronicity Manifesto

Living in flow is a way forward in these crucial times when the old approaches to solving problems are no longer bearing fruit.

When we are able to expect synchronicity and feel flow in our lives, then instead of focusing on our regrets for the opportunities we missed we can see how life has supported the things we *did* choose. We might feel a sense of levity in realizing that any choice is supported by the cosmos, so instead of figuring out the right thing to do, our first task is to get clear inside ourselves on what we are aiming for and then develop the courage to start.

When we are able to expect synchronicity and feel flow in our lives, then instead of feeling like we are up against insurmountable obstacles, we might notice the little gaps in the veneer or cracks in the armor that help us to slowly peel back the layers and allow a solution to emerge.

Rather than feeling like our choices don't really matter, we may see how many opportunities there are to live purposefully and shape our world, even in small ways. We might feel disappointed in our situation, but we certainly won't feel unimportant. How can we feel unimportant when we see the world around us unfolding in creative ways in response to our choices? We might feel discouraged, frustrated, or tired, but we can't avoid seeing the empowerment that meaningful history selection puts in our hands.

From that empowerment, the sense of other people as obstacles may turn into a self-reflection on where we might want to adjust or grow. There are many apples throughout the tree. If a person appears to be in our way, with the apple tree in our minds we might be more able to expand our sense of ourselves and ask if there are other ways we can both be accommodated.

Most importantly of all, although synchronicity shows up in the world outside of us, the more it appears the less we may feel the need to look outside ourselves for meaning and direction. We may come to trust ourselves more.

The only way to know what a meaningful coincidence really means for you is to check inside yourself. Synchronicity is personal. It is nature's language for communicating directly with you. As you develop your personal sense of the meaning of events in your life, you enter into a cosmic dance with the divine that is totally unique to you.

DEBUNKING SYNCHRONICITY

If there is such a phenomenon as synchronicity, it certainly is not new. Throughout history it would have likely influenced the adoption of religions and other forms of prescientific thinking. More recently, the social sciences have done significant work to understand our human tendency to find meaning and patterns in the world, even when they don't objectively exist. It behooves us, then, to examine some of this research and the claims that it makes about synchronicity.

The term "apophenia" describes a tendency to see meaningful connections between unrelated things, which is clearly a consideration when scrutinizing the theory of meaningful history selection I have presented here. The effects I will discuss here are all subcategories of apophenia. If meaningful history selection is correct, then apophenia may not only be a simple case of improperly ascribed meaning.[135] Rather, there may be, in some circumstances, a valid degree of correlation between certain external events and our internal expectation of those events. From this perspective, when psychological confusion arises in relationship to pattern making, an important element to consider might be an inability to internally manage these valid experiences of self-reference in a constructive manner.

How do we generally explain the occurrence of meaningful coincidences? Research suggests that meaningful coincidences are at least partially explained as a cognitive illusion. The argument is that the brain is designed by evolution to be excellent at pattern matching, and it often has a strong motive to form meaningful patterns between events where none may exist. Maybe we are under the influence of strong emotions, such as when we are mad at our spouse for an action that may not actually be connected to past affronts. We might obtain a significant benefit from matching the patterns, leading us to develop superstitions while trying to predict the stock market. Or we might experience a desired sense of safety if we feel like we're connected to the universe and part of a bigger plan. In such situations, our minds postulate patterns in the external world by using cognitive biases, such as affirmation bias, where an individual selectively interprets information to support their preexisting conclusions, or the frequency illusion, where an individual is more likely to notice examples of something they have recently learned about, and they ascribe meaning to its recurrence.

These are all reasonable points that I agree with. The point of contention I will address here is that these arguments are designed to explain the experimental data we have by appealing only to neuroscience, cognitive science, and statistics, because physical science has not yet appeared to produce a good explanation for apophenia. My approach has been somewhat different. I began with a pure investigation into the foundations of quantum mechanics and special relativity, and I found what appeared to be a rigorous connection between what we know in science and the experiences that people have of meaningful coincidence. I am not looking for an *explanation* for synchronicity as much as I am describing what I think the physics says and then looking around to find evidence of it. The phenomenon appears to show up in the form of synchronicity and seems to be rather common. We no longer have to resort to a choice between the supernatural or cognitive illusion, for I believe there is a decent explanation within the realm of physics.

I suspect that a significant number of instances of perceived coincidences are in fact a result of cognitive bias, yet this doesn't mean that all coincidences are illusions of this sort. It seems most likely to me that cognitive bias and meaningful history selection both play a role: meaningful history selection indicates that the cosmos actually does create patterns in response to our actions in the world, and our brains also invent illusory patterns in the world. These two phenomena don't always coincide.

Hence, it seems important to take the phenomenon seriously, in that I suspect meaningful coincidences do happen at a rate greater than chance; but we shouldn't take it too seriously, because we can never know for sure what the relevance of an event is to us. I have emphasized that our ultimate guide is our inner sense of what is right or wrong for us. Rather than painting the situation with a broad brush of reductionist cognitive theory—the conclusion that all apophenia patterns are illusions—we should carefully distinguish between cases of illusion and cases of real physical effects. In other words, I think we have a brain wired to notice patterns, and we also live in a responsive cosmos wired to produce patterns.

Psychologist Richard Wiseman has done research on the phenomenon of luck to determine whether some individuals can be fundamentally luckier than others and whether there are things we can do to become luckier.[136] His findings indicate that to at least some extent, having good luck is a result of habits or behaviors we can adopt that make us more likely to see and act upon surprising opportunities:

> *Unlucky people miss chance opportunities because they are too focused on looking for something else. They go to parties intent on finding their perfect partner and so miss opportunities to make good friends. They look through newspapers determined to find certain types of job advertisements and as a result miss other types of jobs. Lucky people are more relaxed and open, and therefore see what is there rather than just what they are looking for.*[137]

Creating and noticing chance opportunities is the first of four skills that Wiseman says self-described "lucky" people often employ. Lucky people also seem to be good at making lucky decisions based upon their intuition, creating self-fulfilling prophecies via positive expectations, and adopting a resilient attitude that transforms bad luck into good. Wiseman is making an argument about the capacities of individuals to deal with circumstances in a productive way, none of which says anything about the likelihood of coincidental circumstances actually occurring. It is perfectly consistent for there to exist a phenomenon of meaningful history selection, in which meaningful circumstances arise at rates greater than chance, and for individuals to have cognitive biases that affect their interpretation of the situation. These two situations are not mutually exclusive. People do tend to amplify and dramatize the meaningful connections in their lives, but this doesn't mean there are no real meaningful connections. One might suggest that the psychological explanations are sufficient, so there is no need for a physical explanation. In the text I have attempted to illustrate why I think this perspective is an inadequate explanation of the phenomenon. Regardless of whether the phenomenon is needed to explain psychological behavior, I suspect that a physical explanation based upon fundamental physical laws does exist.

I have talked at length about the crucial role of the individual in noticing or even setting the stage for synchronicity by developing certain qualities, such as boldness, authenticity, living from the heart, and getting into flow. Wiseman concludes that "lucky" people are often "more relaxed and open," qualities I have emphasized as well. He is making no claim about the underlying frequency of meaningful circumstances.

Wiseman's suggestion of following intuition to create self-fulfilling prophecies aligns with my suggestion that by following a cycle of listening, opening, reflecting, releasing, and acting we create a positive feedback cycle in which circumstances align with us. We gradually build symbolic momentum toward a particular type of outcome that wouldn't have been very likely without our actions, based on our anticipated experiences, which might be related to Wiseman's "intuition." Wiseman's "positive expectations" are a crucial part of this cycle, because it is only through positive expectations and a resilient attitude that we can interpret the world's responses from a constructive perspective and use that information to take further constructive action, leading to the self-fulfilling prophecy of good luck. Meaningful history selection plays the role of biasing the outcome at each step to align with the choices we've made (based, one hopes, on positive expectations).

At the very least, these perspectives do not appear to conflict. At best, they may be quite well aligned with each other. It seems to me that Wiseman has identified one factor influencing the experience of positive meaningful coincidences, but he assumes it is the only factor. I am not aware of controlled experiments referenced by his work with findings that mitigate against an enhanced frequency of actual meaningful coincidences. Such an experiment would seem quite difficult to conduct, given that these are human beings immersed in the real world, not electrons flying through a slit in a laboratory. He has only made measurements confirming the validity of an explanation for how psychological factors can create or enhance "luck." He appears to conclude that the greater rate of positive meaningful coincidences is a result of selective perception alone. His data, on the other hand, does not appear to indicate that selective perception is the *sole* factor explaining positive meaningful coincidences, just an appealing one.

Interestingly, as in the model presented here, Wiseman also uses an apple analogy to describe how one can increase luck by trying new things. By continually following the same path in an apple orchard, you gradually find fewer and fewer apples because you've already picked apples on those trees. This is similar to my statement that yesterday's bold action becomes today's default pattern (see chapter 4, "Symbolic Momentum"). In the tree of possibilities, we've already gotten to those apples, and if we keep doing the same thing we will continue to get more of the same. But if we want new experiences—or, in Wiseman's case, better luck—we have to stretch into new territory. In my apple tree analogy, this means taking bold action to identify new apples and head to a different area of the tree filled with those apples. In Wiseman's orchard, this means taking a different route through the orchard. The analogies are not the same; his represents a path through a single version of reality (the orchard), whereas mine represents choices between various possible versions of reality (the branches of the tree). Nonetheless, they have useful similarities.

Wiseman says, "The research is ... about encouraging people to move away from a magical way of thinking and toward a more rational view of luck ... [and] using science and skepticism to increase the level of luck, happiness, and success in people's lives."[138] I share this view. The research I've presented goes some distance toward showing that some of what Wiseman would likely consider "magical thinking" can actually be described by a physical mechanism. In other words, it takes a portion of what some in science might call "metaphysics" and proposes to promote it to physics by providing a falsifiable explanation. Wiseman's work does not seem to contradict this approach.

Another commonly cited explanation of synchronicity is the frequency illusion we mentioned earlier.[139] If one is exposed to a new idea or type of object they will find themselves seeing it again and again in the world, which can be at least partially attributed to selective attention (the tendency for our brains to notice things it expects or wants to see).

My approach to synchronicity is to focus on experiences that carry real meaning for an individual, based on the spread of possible outcomes from the actions they can take. In contrast, the frequency illusion is often used to explain circumstances that are quite meaningless or irrelevant to the purposes of the individual.

As an example, let's say we are considering buying a new red sports car. Over the next couple of days we see red cars everywhere: on the freeway, at our friend's house, and in advertisements. We might be tempted to think of this as a "message" confirming that we should purchase the red car. However, in most cases the question of which color car to purchase is totally meaningless, so it is (in my opinion) a waste of time to try to decide whether there is some meaning in seeing the red car everywhere. Treating an example such as this as a valid synchronicity does the field an injustice, because this is more of a parlor trick than a real example of a meaningful phenomenon. If one uses this as an example of synchronicity and then explains it away using a cognitive bias such as the frequency illusion, one might feel that they have "explained" synchronicity. But I believe this is a result of not submitting the matter to careful scientific analysis. One should pick examples that demonstrate the full range of possible meaningful experiences. For instance, if we want to discuss the frequency illusion, I would prefer to use an example such as choosing whether to get married[140] and suddenly encountering references to marriage everywhere you turn.

However, my approach to meaning in coincidences is really just a preference of mine. It is possible that the red car decision is truly meaningful for that individual in that situation, in which case meaningful history selection should apply. It should then be remembered that meaningful history selection supposedly amplifies the likelihood of experiencing circumstances we are anticipating (the red car). It reflects back to us our own actions. Hence, seeing red cars everywhere (or coming across references to marriage) should not be construed as outside advice. It is rather an amplification of the likelihood of seeing those symbols in the world. The meaning or relevance is, in my view, something that comes from inside ourselves. The outer reflection reminds us to check into our own internal guidance and decide whether the circumstance is relevant to our purposes.

Blogger Alan Bellows says the frequency illusion "bears some similarity to synchronicity.... Both phenomena invoke a feeling of mild surprise, and cause one to ponder the odds of such an intersection. Both smack of destiny, as though the events were supposed to occur in just that arrangement ... as though we're witnessing yet another domino tip over in a chain of dominoes beyond our reckoning."[141] My suspicion is that these events are reflections of our inner motivation, not a cosmic message. Therefore, there is a real phenomenon leading to these events, but a mistake can be made in ascribing the origin of the message to a consciousness outside ourselves. In other words, I suggest the problem is not in assigning meaning to external events but in judging the ultimate source as external.

In the same article Bellows suggests that "over the centuries, science has told us that intuition itself is highly flawed, and not to be blindly trusted. The reason for this is our brains' prejudice towards patterns. Our brains are fantastic pattern recognition engines, a characteristic which is highly useful for learning, but it does cause the brain to lend excessive importance to unremarkable events." I agree with this statement from a literal perspective, yet the wording might cause confusion leading to a bias against intuition. The author does not say intuition itself is *completely* flawed and not to be trusted *at all,* yet that seems to be closer to the intended meaning of the sentence; it is certainly the message I obtain from reading it. If so, Bellows is inserting his metaphysical preference to influence the conclusion of the reader. I would agree with a slightly adjusted statement: "Intuition is sometimes unclear and can sometimes lead to inaccurate conclusions about the actual nature of circumstances. Intuition cannot be blindly trusted, yet it may have some benefit; hence, one is well-advised to find an appropriate balance of reason and intuition."

Once again, the frequency illusion can be at least partially explained by cognitive biases, yet this does not lead us to logically conclude that this is the only factor explaining it. The frequency illusion is perfectly consistent with the possibility that physical circumstances can be influenced by our actions through meaningful history selection, if indeed the physics turns out to be correct.

Often we seem to come across authors or individuals who take a somewhat lofty stance and assume that cognitive biases, because they exist, can be used to explain away the phenomenon of synchronicity as an illusion, even going so far as to name the effects as "illusions" rather than "biases." In my opinion these approaches are often overgeneralized. For instance, a person might ask why we need a physical description of synchronicity when these experiences are perfectly within the bounds of statistics and cognitive biases. Yet there are many different types of meaningful experience, and to do science properly we must account for the distinctions between them. Some meaningful coincidences are doubtless explained by statistics and cognitive bias, but this doesn't imply that all of them are. Rather than falling to one or the other metaphysical extreme—either "there are no accidents and everything is synchronicity," or "everything is ultimately described by chance, and meaning is purely subjective"—we should carefully separate out the different instances from each other and recognize when and where multiple factors may influence the likelihood of meaningful coincidences.

INTERPRETATIONS OF QUANTUM MECHANICS

Quantum mechanics is not a clearly understood subject. At its foundation lies ambiguity about what it is supposed to mean about the actual world. Is the "world out there" real? Are interactions between objects limited to objects within causal contact (i.e., objects that are "local" to one another)? Although the mathematical predictions (in the form of statistical trends) have been tested with precision, some underlying questions about the world—and the worldview we should take—are not answered by the theory as it currently stands. Here I will open a window into this debate just large enough for us to see how the issues are relevant for the topics covered in the book.

The general issue is called the measurement problem. It is a reflection of the experimental fact that the state of a system one might measure is a reflection of both the way one measures it and of the system itself. Hence, we must to some degree account for the person or thing doing the measurement.

Furthermore, the math says that any object, when not being observed, should gradually enter into many possible states, like the tree of possibilities presented here. But according to common sense, this doesn't seem to be how normal objects—chairs, people, the moon—behave. So we are left trying to understand why macroscopic objects (i.e., objects that are large enough to see) appear "normally" for us. In the Schrödinger's cat thought experiment, a person puts his cat inside a box such that he can gain no further information about the cat. Then he also places a means of killing the cat in the box—say, a vial of poison—wired to a machine that only gets triggered if, say, a radioactive uranium atom decays. Now, we know that radioactive decay is a quantum process that is described by a branching tree of possibilities. Therefore, on the branches where the radioactive decay happens, the poison is released and the cat dies, and on the other branches it does not. We must logically conclude that even big objects like cats can be described by branches on the tree of possibilities.

So why don't cats really appear like this in real life, both alive and dead at the same time? In my view, this question reveals an unfortunate misunderstanding. A lot of effort has gone into trying to explain this intuitively obvious "fact," which is nevertheless not a fact at all. The reason is that one should never expect to see a cat in a quantum superposition of possibilities, because a quantum superposition occurs only when we are not observing the cat. Hence, any interaction we have with the cat, or the rest of the world for that matter, will always provide a definite result. We will always see a "classical" world, even though that world is purely quantum, because that's exactly what "seeing" does. The same is true of the radioactive atom: if we interact with it, it always takes a definite state, so we never see quantum superposition of microscopic objects either; we only infer it from statistical data. Quantum mechanics is the study of what the world is doing when we are not looking.

Yet much work has gone into explaining what is essentially a metaphysical perspective rather than a scientific one. The mainstream view makes propositions about the macroscopic world: it is both definite (the cat is in a single state) and objective (the "world out there" exists without defining who is observing it). The proposition that the macroscopic world is definite is a metaphysical assumption that tends to line up with people's preferences for how they think the world should be. The proposition that the macroscopic world is objective has been shown quite convincingly to be false, even in the mainstream.[142] Hence it seems to me that this view is supported by neither theory nor fact.

The view that is currently most widely accepted to explain the normal appearance of the macroscopic world is called "decoherence theory." Here's how it works. In chapter 7 we discussed a mirror on a satellite in space that bounces a light particle from the Andromeda galaxy to either a lab in Houston or a lab in Chicago. You might think to yourself that when the light reflected off the mirror in the satellite, or even when it bounced off dust in space long before it came to the satellite, it was "observed" by the satellite or the dust, at which point its possible histories collapsed to just one reality. This is the view taken by decoherence theory. The claim is that everything in the universe acts as an "observer," so that these quantum possibilities can't stay possibilities for long. To establish this claim, however, decoherence theory assumes that the "environment" doesn't follow the rules of quantum mechanics, or more accurately that the strange aspects of quantum mechanics get washed out statistically by very large groups of particles. In other words, the light doesn't exist as a tree of possible states—it is just a single definite state. In the setup to this proposal, the proposal's leading author says "we shall also feel free to assume that a [definite] environment exists,"[143] where the word "definite" in brackets is my addition, emphasizing that he appears to be taking what I call the "objective-definite" nature of the universe for granted.

A large part of my research has to do with supporting an alternate view called "relationality." In a relational universe, decoherence still occurs, but only as a relative process between entities. When two objects interact, they quickly become correlated with each other, and decoherence describes this process; but relative to some other object, no decoherence has occurred. Relationality says, sure, satellites and space dust do observe the light, but they do not erase all the possibilities. Instead, they enter into the possibilities themselves. If you can't beat them, join them! In other words, the tree of possibilities is described relative to each observer.

The timelessness of light (described in chapter 7) leads us to the relational view. Because the beginning point and end point of light's journey are always one single event, and the future where the light gets reflected either toward Chicago or Houston depends on free human choice, then the state that the light is in must depend on who or what is measuring it. For instance, picture the moment *after* the light reflects off the satellite and *before* it arrives on Earth, while it is still travelling through Earth's atmosphere. From the perspective of the satellite, the decision has already been made. But for a lab technician in Houston, waiting to see if they detect the light, the state of the light is not yet established. For how could it be established yet? The light, when it left Andromeda, was (in some of its possibilities) timelessly connected to the detector in their lab! The creation of the light and the detection in Houston together constitute a single (possible) event. There is no way an "intermediate" event, like the choice made at the satellite, could change possibilities that exist simultaneously in the past (light leaving Andromeda) and the future (light detected in Houston).

This is a difficult topic I have taken up elsewhere.[144] In this text we have accordingly assumed relationality to be true and that "Schrödinger's cat" states describe the entire world as we know it. Yet whenever we look at something, those many possibilities always give us a definite, apparently nonquantum single version of that thing. If this were to turn out not to be the case, then alas, the premise of the book would be faulty.

Also controversial is my premise that qualitative experiences or qualia are the fundamental building blocks of nature. This topic is researched in philosophy and cognitive neuroscience, both outside my own expertise. Proponents of this view of qualia include philosopher David Chalmers, whose doctoral dissertation defined the problem for a whole generation of researchers.[145] Critics of this view, such as philosopher Daniel Dennett, claim that qualia are fundamentally unnecessary to describe reality because qualitative experience can be explained in terms of still more fundamental physical properties described in physics, chemistry, and biology.[146]

While some see this debate as purely philosophical, I suspect that it leads to measurable phenomena such as synchronicity and can therefore eventually be tested and entered into the domain of physics.

CALCULATING THE ODDS OF A SYNCHRONICITY

In chapter 4, I described a synchronicity in which I attended a spiritual service for four weeks in a row, hoping for an opportunity to volunteer as a musician. After the fifth week I received an unrelated phone call offering me a paid job as a musical director for a similar spiritual center. Here I will discuss in brief how one might prove that this is indeed a meaningful coincidence and not just a fluke.

I do not yet know how to confidently calculate the odds of experiences such as this one, with so many factors involved. We might ask what the likelihood is that I would receive the job offer on that particular week out of the entire year of weeks surrounding it, during which I was actively working to build my music career. That chance is about one out of fifty, or 2 percent. This is rather low, below the 5 percent threshold at which we tend to think another explanation may be needed, but this is a very rough estimate. Also, it was bound to happen some week, so such a simple statistical argument is quite weak.

Let's try a slightly more nuanced approach. It is actually fairly reasonable that someone like Reverend Mary would find my name and seek me out, but what is truly serendipitous here is the timing of her call. To get a sense of why this a meaningful coincidence, let's consider that I had been playing music professionally at that point for two and a half years, or roughly 120 weeks. Of that time, only in the recent six weeks or so had I acted on my intention to become part of this community. The likelihood of her calling at that precise window of time is about six out of 120, or 5 percent, which is again the general cutoff for determining whether an experimental result is best described by chance.

I suspect the numbers become more convincing when we look at the ubiquity of meaningful coincidences like this in daily life and calculate the odds that they would *all* turn out this way. In other words, if every potential synchronicity in my day has a likelihood of around 5 percent, it would seem pretty unlikely that I would have more than one or two in a day. Yet if one defines synchronicity the way I have, as a coincidence that is at least slightly meaningful, then these events are quite commonplace.

It is not clear how one would approach an in-depth analysis of this. Joseph Mazur's book *Fluke*[147] takes some interesting steps toward addressing this type of calculation. Still, I don't think the necessary statistical calculations have yet been done to accurately describe a situation like the one here. Some have argued that due to selective attention, we are not accurately accounting for all the "near misses" in a day, meaning that we actually experience very few of all the possible synchronicities. Similarly, if a day has 1,440 minutes in it, and a given synchronicity takes up only five minutes, that leaves 287 other five-minute blocks in the day during which no meaningful coincidence happened, so it seems that we are 287 times more likely to *not* receive a synchronicity.

Yet I believe this is misleading. The number of *salient* events in a day is much smaller. A given day only has a handful of salient or significant events. When we say "I experienced a synchronicity on my commute to work," we are pointing out that the entire commute was essentially one salient moment. So in that case our salient block of time is not five minutes but, say, twenty-five minutes.

When considering synchronicity, time is not broken up in equal-size chunks in every situation. Rather, different situations will unfold over different lengths of time. Showing up to a performance at the symphony is really just a couple of salient moments, although the entire symphony might be three hours long. There might be a synchronicity when you stand in line for the ticket booth or the restrooms and see a dear friend, or maybe the program includes pieces of music you have recently learned, or perhaps you find you are seated next to the composer of the piece, whose work you have studied fervently.

In general, a single day is composed of only so many "moments" like these when something meaningful can happen. The branchings of the tree are not measured by ticks of a clock but by relevant interactions, which can vary widely depending on the many factors present in a normal life situation. Rather than seeing time from a reductionist perspective—one in which every second counts as a possibility for synchronicity (and most of the seconds don't have synchronicity)—I think we can look at time as a collection of salient moments, of which there are only about twenty or so a day.

The following comparison gives us another way to understand this. In the reductionist view, if I wait in line at the symphony for twenty minutes and I coincidentally meet someone who was just hired by my company and is starting next week, we might call that a synchronicity. Let's say our interaction is only a minute long. Does that mean there are nineteen other minutes in which a synchronicity didn't happen? That's a very unreasonable perspective that would lead one to conclude that 95 percent of the moments in our day are "misses," with the occasional synchronicity "hit." Instead, a more reasonable view is that the entire twenty-minute wait in line is *one* moment, or *one* setup, during which *one* synchronicity happened. From an experimental perspective, this entire moment would be considered a "hit."

Thus, if we have more than a few meaningful coincidences in a day, which I suspect is not uncommon, and if the likelihood of each one is on the order of 2 percent, the series of them together is quite unlikely. For instance, three meaningful coincidences per day, each of which has a likelihood of 2 percent, would have a likelihood somewhere on the order of 0.0008 percent. Admittedly, this argument relies on a number of assumptions that I cannot justify here, but the relevant point is that one should look at the likelihood of an *ongoing series of small synchronicities* rather than just the likelihood of a single, isolated, dramatic synchronicity.

More academic work on this question would be beneficial to the field.

GLOSSARY

Most of the terms in this glossary are not necessary to understand most of the ideas in the text. They are provided here as reference for those readers who want to dig deeply into the scientific concepts discussed.

amplitude: A number that measures probability and that relates to how heavy a branch of the tree of possibilities is. If we square the probability amplitude, we get the likelihood of that branch occurring for the object.

anticipated qualitative experience: It is proposed in the text that when we are motivated to act in the world, we are always "anticipating" a future qualitative experience (see "qualitative experience," below) that is either the same as or different from the present one. This natural capacity to imagine and feel an intended experience is the driving factor in meaningful history selection.

consistency: Although the properties of objects are undefined before we measure them, they cannot take just any value whatsoever. Rather, they are constrained to be consistent with (or agree with) any measurements you have already made that the object has been affected by.

consistent/decoherent histories quantum mechanics: An interpretation of quantum mechanics that focuses on chains of quantum events. Paradoxes in quantum mechanics are resolved by insisting on the single-framework rule, in which we must choose a single perspective from which to view a given situation.

correlated superposition of possibilities: See "quantum correlation," below.

counterfactual indefiniteness: The experimentally verified concept that one cannot speak meaningfully about what would have occurred if a different set of choices had been made. "What would have happened if ..." is a meaningless question.

decoherence theory: A leading theory that describes why the quantum behavior of objects at microscopic scales appears to not apply at macroscopic scales. This book disagrees with that premise.

entanglement: See "quantum correlation," below.

meaningful alignment: If a particular branch on the tree of possibilities represents a configuration of the world in which a specific qualitative experience occurs, and if the observer takes an action motivated by a similar anticipated qualitative experience, then there is a meaningful alignment between the observer's action and the branch.

meaningful branching: An uneven spread of apples on the branches of the tree of possibility; a branch point at which some branches are very likely to lead to a specific type of outcome (lots of apples) and other branches are not likely to do so (very few apples). A meaningful branching is only meaningful with respect to the particular type of outcome (apple) specified. It may be non-meaningful (an even spread of apples) for another type of outcome.

meaningful coincidence: See "synchronicity," below.

objective-definite: An event that is objective-definite would be both objective (true for everybody) and definite (having one and only one set of fixed properties). This book claims that such states don't exist; rather, the cosmos is relational such that for a given observer the measured world is definite, but one cannot take more than one perspective at a time. Therefore one cannot claim to experience an objective result that is also true for observers who have not made a measurement.

qualitative experience: The qualitative aspect of living, such as "the experience of eating a cherry." Some philosophers see the stream of qualitative experiences as the fundamental characteristic of living that each individual can point to as constituting life. Qualitative experiences cannot be transferred or conveyed objectively between two different observers; they are fundamentally subjective yet consistent. (Also "qualia.")

quantum correlation: When two quantum objects have interacted in the past but have not yet been measured by a specific observer, they exist together in a superposition of possible properties for that observer, and their possible properties must match each other.

quantum object: An object that obeys the laws of quantum mechanics, such as superposition and interference. This generally is considered to apply only to microscopic particles, but the research referenced here advocates for the view that all objects, including macroscopic ones, obey these laws.

relational: Having definite properties only relative to a given observer; the opposite of objective-definite. If I measure something, I get a definite result relative to me. If you come along, you find both me and the object in a (correlated) superposition of possibilities. (Also "relationally defined" or "observer-dependent.")

relational quantum mechanics: An interpretation of quantum mechanics which postulates that the results of measurements are personal to the observer, not objective-definite.

serendipity: The same as synchronicity, although typically thought of as specifically beneficial in nature.

singular moment: A branching point on the tree of possibilities whose child branches lead to significantly different outcomes. An important choice point.

superposition: The condition where two or more sets of mutually exclusive circumstances are all still possible. Only one of the circumstances can end up being true when a measurement is made, but before the object is measured we have a superposition of many different potentially true circumstances. Imagine the picture on the wall if a transparency projector displays more than one overlapping transparency at the same time.

synchronicity: Unlikely circumstances that become more likely because they align with an observer's purpose or inner experience. Agnostic in result; can be beneficial or detrimental.

tree of possibilities: A metaphor used to describe the evolution of possibilities as being shaped like a branching tree. One's current situation is generally considered to be at the base of the tree, and each decision or interaction with one's surroundings results in the tree branching further. Each branch has an amplitude or weight associated with it, which corresponds indirectly to the likelihood of that branch occurring.

undetermined states: Before we measure an object, it is in a superposition of possibilities relative to us, so the actual "true" properties of the object are undetermined.

wave function: A mathematical structure describing the spread of possible properties of an object.

NOTES

1. Mazur, *Fluke;* Burger and Starbird, *Coincidences;* Taleb, *Fooled by Randomness;* Hand, *Improbability Principle.*
2. Jung, *Synchronicity,* 21.
3. Eisenstein, *More Beautiful World.*
4. Nelson-Isaacs, "Retroactive Event Determination and the Interpretation of Macroscopic Quantum Superposition States"; Nelson-Isaacs, "Retroactive Event Determination and its Relativistic Roots"; Nelson-Isaacs, "Guiding Quantum Histories."
5. Stephen Gaertner, email message to author, April 13, 2018.
6. Csikszentmihalyi, *Flow: The Psychology of Optimal Experience;* Csikszentmihalyi, *Flow and the Foundations of Positive Psychology;* Csikszentmihalyi and Nakamura, "The Concept of Flow."
7. Csikszentmihalyi, *Flow: The Psychology of Optimal Experience,* 58.
8. Jaworski, *Synchronicity,* 185.
9. Diggins, "The True History."
10. Csikszentmihalyi, *Flow: The Psychology of Optimal Experience,* 59.
11. Jung, *Synchronicity,* 19.
12. Berger and Johnston, *Simple Habits;* Laloux, *Reinventing Organizations;* Patterson et al., *Crucial Conversations;* Merry, "Synchronicity and Leadership."
13. Patterson et al., *Crucial Conversations,* 24.
14. Caprino, "The Top 10 Things."
15. See appendix C for more discussion on this.
16. Jung, *Synchronicity;* Jung et al., *Man and His Symbols.*
17. Peat, *Synchronicity.*
18. Jaworski, *Synchronicity.*
19. Combs and Holland, *Synchronicity.*
20. Merry, "Synchronicity and Leadership."
21. Baets, *Complexity, Learning and Organizations.*

22 Beitman, *Connecting with Coincidence.*
23 Lorenz, "Synchronicity in the 21st Century."
24 Scharmer et al., "Illuminating the Blind Spot."
25 Surprise, *Synchronicity.*
26 Chopra, *Synchrodestiny.*
27 Koestler, *Roots of Coincidence.*
28 Kammerer quoted in Jung, *Synchronicity*, 8.
29 Csikszentmihalyi, *Flow: The Psychology of Optimal Experience*, 58.
30 Csikszentmihalyi, *Flow: The Psychology of Optimal Experience*, 6.
31 Kotler, *Rise of Superman.*
32 Amabile et al., "Affect and Creativity."
33 Limb and Braun, "Neural Substrates."
34 Csikszentmihalyi, *Flow: The Psychology of Optimal Experience.*
35 Juster, *Phantom Tollbooth*, 132.
36 Chalmers, "Toward a Theory of Consciousness"; Nagel, "What Is It Like to Be a Bat?"
37 Jackson, "What Mary Didn't Know."
38 Zukav, *Seat of the Soul*, 107.
39 Jung, "Concept of the Collective Unconscious," 100.
40 Ekman, "Basic Emotions."
41 Damasio, *Feeling of What Happens.*
42 Chalmers, "Toward a Theory of Consciousness"; Chalmers, "Facing Up to the Problem of Consciousness."
43 Zukav, *Seat of the Soul*, 39.
44 Damasio, *Feeling of What Happens*, 136.
45 Patterson et al., *Crucial Conversations*, 106.
46 Fyfe, "Lithium and Bipolar Disorder."
47 Jung, *Synchronicity*, 32.
48 Jung, *Synchronicity*, 32.
49 Zukav, *Seat of the Soul*, 39.
50 Wheeler, "Information, Physics, Quantum," 311.
51 Jaworski, *Synchronicity*, 182.
52 McMaster, *Lifeshocks.*
53 Laloux, *Reinventing Organizations.*
54 Laloux, *Reinventing Organizations*, 100.

55 Combs and Holland, *Synchronicity*, 135.
56 Jaworski, *Synchronicity*, 184.
57 Wann, "Examining the Potential Causal Relationship."
58 Laloux, *Reinventing Organizations*.
59 Lent, *Patterning Instinct*.
60 Wann, "Examining the Potential Causal Relationship"; Wann et al., "Examining Sport Team Identification."
61 Andriessen and Krysinska, "Essential Questions."
62 Volk and Lagzdins, "Bullying and Victimization."
63 Peguero and Williams, "Racial and Ethnic Stereotypes."
64 Berger and Johnston, *Simple Habits*.
65 Laloux, *Reinventing Organizations*.
66 PriceWaterhouseCoopers, "Putting Purpose to Work"; Achieve Agency, *Millennial Impact Report*.
67 Patterson et al., *Crucial Conversations*, 24.
68 Rosenberg, *Nonviolent Communication*.
69 Senge et al., *Presence*.
70 Jaworski, *Synchronicity*; Jaworski, *Source*.
71 Jaworski, *Source*, 182.
72 Scharmer, *Theory U*.
73 Berger and Johnston, *Simple Habits*, 54.
74 Patterson et al., *Crucial Conversations*, 53–58.
75 Merry, "Synchronicity and Leadership," 187.
76 Kuhn, *Structure*, quoted in Csikszentmihalyi, *Flow: The Psychology of Optimal Experience*, 135. In my opinion, Kuhn's comments apply to both genders, though they explicitly reference the male gender.
77 Csikszentmihalyi, *Finding Flow*, 56.
78 Csikszentmihalyi, *Finding Flow*, 54–55.
79 Wells, "Self-Esteem," quoted in Csikszentmihalyi, *Finding Flow*, 58.
80 See appendix C for a discussion of these odds.
81 Jaworski, *Synchronicity*, 184.
82 Csikszentmihalyi, *Flow: The Psychology of Optimal Experience*, 3.
83 Jung, *Archetypes*; Jung, *Two Essays*. Carl Jung's research on archetypes may be a useful framework for future research on symbolic meaning.
84 Lent, *Patterning Instinct*.

85 Combs and Holland, *Synchronicity*, 133.
86 Hoffman, Singh, and Prakash, "Interface Theory."
87 Hameroff, "Quantum Origin."
88 Wiseman, "Luck Factor."
89 Tzu, *Tao Te Ching*, 48.
90 Zukav, *Seat of the Soul*, 44.
91 Van Ness and Strong, *Restoring Justice*.
92 Rudd, Vohs, and Aaker, "Awe Expands People's Perception of Time"; Piff et al., "Awe"; Armenta, Fritz, and Lyubomirsky, "Functions of Positive Emotions."
93 Armenta, Fritz, and Lyubomirsky, "Functions of Positive Emotions," 2.
94 Plath, *Bell Jar*, 85–86.
95 Csikszentmihalyi, *Flow: The Psychology of Optimal Experience*, 7.
96 Armenta, Fritz, and Lyubomirsky, "Functions of Positive Emotions," 6.
97 The way I have defined them here, ego habits appear to be related to samskaras in Hindu philosophy, as well as the defense mechanisms of Freudian analysis. These fields lie outside my expertise, so I will not attempt to clarify the connection, but I will speak from personal experience as well as basic understanding of the relevant principles of yoga and psychoanalysis. See Satchidananda, *Yoga Sutras*; Freud, *Civilization*; Freud, *Ego*.
98 Don't tell my editor.
99 Zukav, *Seat of the Soul*, 145–46.
100 Combs and Holland, *Synchronicity*, 136.
101 Porter, "Why You Should Make Time"; Di Stefano et al., "Making Experience Count"; Jachimowicz et al., *Commuting as Role Transitions*.
102 Cheung et al., "People with High Self-Control."
103 Skipper, "Secret to Being a Productive Human."
104 Csikszentmihalyi, *Creativity*, 76.
105 Csikszentmihalyi, *Finding Flow*, 57.
106 Csikszentmihalyi, *Flow: The Psychology of Optimal Experience*, 44.
107 Diener, Horwitz, and Emmons, "Happiness."
108 Heroic Imagination Project.
109 Bohm, *Wholeness*, 163.

Notes

110 See appendix A for a discussion of Einstein's perspective on this.
111 Griffiths, *Consistent Quantum Theory;* Rovelli, "Relational Quantum Mechanics"; Mermin, "Ithaca Interpretation"; Aharonov and Vaidman, "Two-State Vector Formalism"; Everett, "'Relative State' Formulation."
112 Nelson-Isaacs, "Retroactive Event Determination and its Relativistic Roots"; Nelson-Isaacs, "Retroactive Event Determination and the Interpretation of Macroscopic Quantum Superposition States."
113 Gehrenbeck, "Electron Diffraction."
114 Jung, *Synchronicity,* 35.
115 Heisenberg, *Physics and Philosophy,* 52.
116 As quoted in Donati, "Beyond Synchronicity," 715.
117 Mermin, "What's Wrong with This Pillow?"; Lahiri, "Shut Up."
118 Nelson-Isaacs, "Retroactive Event Determination and its Relativistic Roots"; Nelson-Isaacs, "Fourier Kinematics and the Implicate Order."
119 Bohm, *Wholeness,* 191.
120 Bohm, *Wholeness,* 188.
121 Bohm, *Wholeness,* 212.
122 Bohm, *Wholeness,* 76.
123 Rovelli, "Relational Quantum Mechanics."
124 See appendix B for a discussion of Schrödinger's cat.
125 See appendix B for a discussion of macroscopic quantum superposition states.
126 Reiher, "Experiences with Optimistic Synchronization."
127 Reiher, "Experiences with Optimistic Synchronization," 4.
128 Monneret, "Platinum Anticancer Drugs."
129 Jung, *Synchronicity,* 29–30.
130 Zukav, *Seat of the Soul,* 170.
131 Jung, *Synchronicity.*
132 Jaworski, *Synchronicity,* 182.
133 Jaworski, *Synchronicity,* 182.
134 Laloux, *Reinventing Organizations,* 48.
135 Mishara, "Klaus Conrad."
136 Wiseman, "Luck Factor."

137 Wiseman, "Luck Factor."
138 Wiseman, "Luck Factor."
139 Zwicky, "More Illusions."
140 Merry, "Synchronicity and Leadership," 12.
141 Bellows, "Baader-Meinhof."
142 Mermin, "What's Wrong with This Pillow?"; Mermin, "What Is Quantum Mechanics Trying to Tell Us?"; Kochen and Specker, "Problem of Hidden Variables"; Bell, "On the Problem of Hidden Variables"; Hensen et al., "Loophole-Free Bell Inequality Violation."
143 Zurek, "Decoherence," 4.
144 Nelson-Isaacs, "Retroactive Event Determination and Its Relativistic Roots"; Nelson-Isaacs, "Retroactive Event Determination and the Interpretation of Macroscopic Quantum Superposition States."
145 Chalmers, "Toward a Theory of Consciousness."
146 Dennett, "Consciousness Explained."
147 Mazur, *Fluke*.

BIBLIOGRAPHY

Achieve Agency. *The Millennial Impact Report.* West Palm Beach: Achieve Agency, 2017. www.themillennialimpact.com/.

Aharonov, Yakir, and Lev Vaidman. "The Two-State Vector Formalism: An Updated Review." In *Time in Quantum Mechanics—Vol. I,* edited by Gonzalo Muga, R. S. Mayato, and Íñigo Egusquiza, 399–447. Lecture Notes in Physics 734. Berlin: Springer-Verlag, 2008.

Amabile, Teresa M., Sigal G. Barsade, Jennifer S. Mueller, and Barry M. Staw. "Affect and Creativity at Work." *Administrative Science Quarterly* 50, no. 3 (2016): 367–403.

Andriessen, Karl, and K. Krysinska. "Essential Questions on Suicide Bereavement and Postvention." *Int J Environ Res Public Health* 9, no. 1 (2012): 24–32.

Armenta, Christina N., Megan M. Fritz, and Sonja Lyubomirsky. "Functions of Positive Emotions: Gratitude as a Motivator of Self-Improvement and Positive Change." *Emotion Review* 9, no. 3 (2017): 183–190. doi.org/10.1177/1754073916669596.

Baets, Walter. *Complexity, Learning and Organizations: A Quantum Interpretation of Business.* New York: Routledge, 2006.

Beitman, Bernard. *Connecting with Coincidence: The New Science for Using Synchronicity and Serendipity in Your Life.* Deerfield Beach: Health Communications, 2016.

Bell, John S. "On the Problem of Hidden Variables in Quantum Mechanics." *Reviews of Modern Physics* 38, no. 3 (1966): 447–52.

Bellows, Alan. "The Baader-Meinhof Phenomenon." *Damn Interesting,* March 19, 2006. www.damninteresting.com/the-baader-meinhof-phenomenon/.

Berger, Jennifer G., and Keith Johnston. *Simple Habits for Complex Times.* Stanford: Stanford University Press, 2015.

Bohm, David. *Wholeness and the Implicate Order.* London: Routledge, 1980.

Burger, Edward B., and Michael Starbird. *Coincidences, Chaos, and All That Math Jazz: Making Light of Weighty Ideas.* New York: W. W. Norton & Company, 2005.

Caprino, Kathy. "The Top 10 Things People Want in Life But Can't Seem to Get." *Huffington Post.* March 29, 2016. www.huffingtonpost.com/kathy-caprino/the-top-10-things-people-_2_b_9564982.html.

Chalmers, David. "Facing Up to the Problem of Consciousness." *Journal of Consciousness Studies* 2 , no. 3 (1995): 200–219. consc.net/papers/facing.html.

Chalmers, David. "Toward a Theory of Consciousness." PhD diss., Indiana University, 1993.

Cheung, Tracy T. L., Marleen Gillebaart, Floor Kroese, and Denise De Ridder. "Why Are People with High Self-Control Happier? The Effect of Trait Self-Control on Happiness as Mediated by Regulatory Focus." *Front Psychol* 5 (722), 2014. doi.org/10.3389/fpsyg.2014.00722.

Chopra, Deepak. *Synchrodestiny: Harnessing the Infinite Power of Coincidence to Create Miracles.* London: Rider, 2004.

Combs, Alan, and Mark Holland. *Synchronicity: Through the Eyes of Science, Myth and the Trickster.* New York: Marlowe & Company, 1996.

Csikszentmihalyi, Mihaly. *Creativity: Flow and the Psychology of Discovery and Invention.* New York: HarperCollins, 1996.

———. *Finding Flow: The Psychology of Engagement with Everyday Life.* New York: Basic Books, 1997.

———. *Flow and the Foundations of Positive Psychology: The Collected Works of Mihaly Csikszentmihalyi.* New York: Springer, 2014.

———. *Flow: The Psychology of Optimal Experience.* New York: Harper & Row, 1990.

——— and Jeanne Nakamura. "The Concept of Flow." In *Oxford Handbook of Positive Psychology,* edited by Shane Lopez and C. R. Snyder, 89–105. New York: Oxford University Press, 2009.

Damasio, Antonio. *The Feeling of What Happens: Body and Emotion in the Making of Consciousness.* Wilmington, MA: Mariner Books, 2000.

Dennett, Daniel C. *Consciousness Explained.* New York: Little, Brown and Co., 1991.

Diener, Ed, Jeff Horwitz, and Robert A. Emmons. "Happiness of the Very Wealthy." *Social Indicators Research* 16, no. 3 (1985): 263–74.

Diggins, F. W. E. "The True History of the Discovery of Penicillin, with Refutation of the Misinformation in the Literature." *British Journal of Biomedical Science* 56, no. 2 (1999): 83–93.

Di Stefano, Giada, Francesca Gino, Gary P. Pisano, and Bradley Staats. "Making Experience Count: The Role of Reflection in Individual Learning." Harvard Business School NOM Unit working paper no. 14-093; Harvard Business School Technology and Operations Management Unit working paper no. 14-093; HEC Paris research paper no. SPE-2016-1181. June 14, 2016. papers.ssrn.com/sol3/papers.cfm?abstract_id=2414478.

Donati, Marialuisa. "Beyond Synchronicity: The Worldview of Carl Gustav Jung and Wolfgang Pauli." *Journal of Analytical Psychology* 49, no. 5 (2004): 707–28.

Easwaran, Eknath. *Gandhi the Man: The Story of His Transformation.* 3rd ed. Berkeley: Blue Mountain Center of Meditation, 1997.

Ekman, Paul. "Basic Emotions." In *Handbook of Cognition and Emotion*, edited by Mick Power and Tim Dalgleish, 45–60. West Sussex: Wiley, 1999. www.paulekman.com/wp-content/uploads/2013/07/Basic-Emotions.pdf.

Eisenstein, Charles. *The More Beautiful World Our Hearts Know Is Possible.* Berkeley: North Atlantic Books, 2013.

Everett, Hugh, III. "'Relative State' Formulation of Quantum Mechanics." *Reviews of Modern Physics* 29, no. 3 (1957): 454–62. journals.aps.org/rmp/abstract/10.1103/RevModPhys.29.454.

Freud, Anna. *The Ego and the Mechanisms of Defence.* London: Hogarth Press, 1937.

Freud, Sigmund. *Civilization and Its Discontents.* New York: Jonathan Cape & Harrison Smith, 1930.

Fyfe, Ian. "Lithium and Bipolar Disorder—Exploiting the Unexpected." *The Biochemist* 31, no. 6 (2009): 4–6.

Gehrenbeck, Richard K. "Electron Diffraction: Fifty Years Ago." *Physics Today* 31, no. 1 (1978): 34–41. doi.org/10.1063/1.3001830.

Griffiths, Robert B. *Consistent Quantum Theory.* New York: Cambridge University Press, 2002.

Hameroff, Stuart. "The Quantum Origin of Life: How the Brain Evolved to Feel Good." In *On Human Nature: Biology, Psychology, Ethics, Politics, and Religion*, edited by Michel Tibayrenc and Francisco J. Ayala, 333–53. Tucson: Elsevier, 2016. www.sciencedirect.com/science/article/pii/B978012420190300020X.

Hand, David J. *The Improbability Principle: Why Coincidences, Miracles, and Rare Events Happen Every Day*. New York: Scientific American/Farrar, Straus and Giroux, 2014.

Heisenberg, Werner. *Physics and Philosophy: The Revolution in Modern Science*. New York: Harper Perennial Modern Classics, 2007.

Hensen, B., H. Bernien, A. E. Dréau, A. Reiserer, N. Kalb, M. S. Blok, J. Ruitenberg, et al. "Loophole-Free Bell Inequality Violation Using Electron Spins Separated by 1.3 Kilometres." *Nature* 526 (2015): 682–86. dx.doi.org/10.1038/nature15759.

The Heroic Imagination Project. www.heroicimagination.org/.

Hoffman, Donald D., Manish Singh, and Chetan Prakash. "The Interface Theory of Perception." *Psychon Bull Rev* 22, no. 6 (2015): 1480–506. doi:10.3758/s13423-015-0890-8.

Jachimowicz, Jon M., Julia J. Lee, Bradley R. Staats, Jochen I. Menges, and Francesca Gino. *Commuting as Role Transitions: How Trait Self-Control and Work-Related Prospection Offset Negative Effects of Lengthy Commutes*. Working Paper 16-077, Harvard Business School. 2016. www.hbs.edu/faculty/Publication%20Files/16-077_0aae29b7-b67e-402a-9bc9-c06dd477e8da.pdf.

Jackson, Frank. "What Mary Didn't Know." *Journal of Philosophy* 83, no. 5 (1986): 291–95.

Jaworski, Joseph. *Source: The Inner Path of Knowledge Creation*. San Francisco: Berrett-Koehler Publishers, 2012.

———. *Synchronicity: The Inner Path of Leadership*. San Francisco: Berrett-Koehler Publishers, 2011.

Jung, Carl Gustav. *The Archetypes and the Collective Unconscious*. Vol. 9, part 1 of *Collected Works of C. G. Jung*. Princeton: Bollingen, 1981.

———. *The Concept of the Collective Unconscious: A Lecture Delivered before the Analytical Psychology Club of New York City, October 2, 1936*. New York: The Club, 1936.

———. *Synchronicity: An Acausal Connecting Principle.* Princeton: Princeton University Press, 1952.

———. *Two Essays on Analytical Psychology.* Vol. 7 of *Collected Works of C. G. Jung.* London: Routledge, 1966.

———, Marie-Louise von Franz, Joseph L. Henderson, Aniela Jaffé, and Jolande Jacobi. *Man and His Symbols.* New York: Doubleday, 1964.

Juster, Norton. *The Phantom Tollbooth.* New York: Random House Children's Books, 1961.

Kochen, Simon, and Ernst P. Specker. "The Problem of Hidden Variables in Quantum Mechanics." *Journal of Mathematics and Mechanics* 17 (1967): 59–87.

Koestler, Arthur. *The Roots of Coincidence.* London: Hutchinson & Co, 1972.

Kotler, Steven. *The Rise of Superman: Decoding the Science of Ultimate Human Performance.* London: Quercus Publishing, 2015.

Kuhn, Thomas. *The Structure of Scientific Revolutions.* Chicago: University of Chicago Press, 1962.

Lahiri, Avijit. "Shut Up and Calculate!" May 13, 2014. www.physicsandmore.net/resources/Shutupandcalculate.pdf.

Laloux, Frederik. *Reinventing Organizations.* Brussels: Nelson Parker, 2014.

Lama, H. H. Dalai, and Howard C. Cutler. *The Art of Happiness, 10th Anniversary Edition: A Handbook for Living.* New York: Riverhead Books, 2009.

Lent, Jeremy. *The Patterning Instinct: A Cultural History of Humanity's Search for Meaning.* Amherst: Prometheus Books, 2017.

Limb, Charles J., and Allen R. Braun. "Neural Substrates of Spontaneous Musical Performance: An fMRI Study of Jazz Improvisation." *PLoS ONE* 3, no. 2 (2008): e1679. doi.org/10.1371/journal.pone.0001679.

Lorenz, Helene Shulman. "Synchronicity in the 21st Century." *Journal of Jungian Scholarly Studies* 2, no. 2 (2006). jungiansociety.org/images/e-journal/Volume-2/Lorenz-2006.pdf.

Mazur, Joseph. *Fluke: The Math and Myth of Coincidence.* New York: Basic Books, 2016.

McMaster, Ann. *Lifeshocks Out of the Blue: Learning from Your Life's Experiences.* Self-published, CreateSpace, 2015.

Mermin, Nathaniel David. "The Ithaca Interpretation of Quantum Mechanics." *Pramana* 51, no. 5 (1998): 549–65. doi.org/10.1007/BF02827447.

———. "What Is Quantum Mechanics Trying to Tell Us?" *American Journal of Physics* 66 (1998): 753–67.

———. "What's Wrong with This Pillow?" *Physics Today* 42, no. 4 (1989). physicstoday.scitation.org/doi/10.1063/1.2810963.

Merry, Philip. "Synchronicity and Leadership." PhD diss., Tilburg University, 2017.

Mishara, Aaron L. "Klaus Conrad (1905–1961): Delusional Mood, Psychosis, and Beginning Schizophrenia." *Schizophrenia Bulletin* 36, no. 1 (2010): 9–13. doi.org/10.1093/schbul/sbp144.

Monneret, C. "Platinum Anticancer Drugs. From Serendipity to Rational Design." *Ann Pharm Fr* 69, no. 6 (2011): 286–95. doi:10.1016/j.pharma.2011.10.001.

Nagel, Thomas. "What Is It Like to Be a Bat?" *Philosophical Review* 83, no. 4 (1974): 435–50.

Nelson-Isaacs, Sky. "Fourier Kinematics and the Implicate Order." Unpublished manuscript, 2018.

———." Guiding Quantum Histories with Intermediate Decomposition of the Identity." *AIP Conference Proceedings* 1841, no. 1 (2017). aip.scitation.org/doi/abs/10.1063/1.4982770.

———. "Retroactive Event Determination and its Relativistic Roots." In *Quantum Retrocausation—Theory and Experiment*, edited by Daniel P. Sheehan, 45–74. San Diego: American Institute of Physics, 2011.

———. "Retroactive Event Determination and the Interpretation of Macroscopic Quantum Superposition States in Consistent Histories and Relational Quantum Mechanics." *Journal for Scientific Exploration* 25, no. 2 (2011): 273–304.

Patterson, Kerry, Joseph Grenny, Ron McMillan, Al Switzler, and Laura Roppe. *Crucial Conversations: Tools for Talking When the Stakes Are High*. New York: McGraw Hill, 2012.

Peat, F. David. *Synchronicity: The Bridge Between Matter and Mind*. New York: Bantam, 1987.

Peguero, Anthony A., and Lisa M. Williams. "Racial and Ethnic Stereotypes and Bullying Victimization." *Youth & Society* 45, no. 4 (2013): 545–64.

Piff, Paul K., Pia Dietze, Matthew Feinberg, Daniel M. Stancato, and Dacher Keltner. "Awe, the Small Self, and Prosocial Behavior." *Journal of Personality and Social Psychology* 108, no. 6 (2015): 883–99. doi:10.1037/pspi0000018.

Plath, Sylvia. *The Bell Jar.* New York: Harper & Row, 1971.

Porter, Jennifer. "Why You Should Make Time for Self-Reflection (Even If You Hate Doing It)." *Harvard Business Review,* March 21, 2017. hbr.org/2017/03/why-you-should-make-time-for-self-reflection-even-if-you-hate-doing-it.

PriceWaterhouseCoopers. "Putting Purpose to Work: A Study of Purpose in the Workplace." June 2016. www.pwc.com/us/en/about-us/corporate-responsibility/assets/pwc-putting-purpose-to-work-purpose-survey-report.pdf.

Reiher, Peter L. "Experiences with Optimistic Synchronization for Distributed Operating Systems." In *Proceedings of the Third Symposium on Experiences with Distributed and Multiprocessor Systems,* 59–78. Newport Beach: USENIX Association, 1992. lasr.cs.ucla.edu/reiher/papers/tw_experiences.pdf.

Rosenberg, Marshall. *Nonviolent Communication: A Language of Life.* Encinitas: PuddleDancer Press, 2003.

Rovelli, Carlo. "Relational Quantum Mechanics." *Int J Theor Phys* 35, no. 8 (1996): 1637–78. doi.org/10.1007/BF02302261.

Rudd, Melanie, Kathleen D. Vohs, and Jennifer Aaker. "Awe Expands People's Perception of Time, Alters Decision Making, and Enhances Well-Being." *Psychological Science* 23, no. 10 (2012): 1130–36. www.jstor.org/stable/23355506.

Satchidananda. *The Yoga Sutras of Patanjali.* Yogaville: Integral Yoga Publications, 1978.

Scharmer, C. Otto. *Theory U: Leading from the Future as It Emerges.* San Francisco: Berrett-Koehler, 2009.

———. W. Brian Arthur, Jonathan Day, Joseph Jaworski, Michael Jung, Ikujiro Nonaka, and Peter M. Senge. "Illuminating the Blind Spot: Leadership in the Context of Emerging Worlds." 2002. www.ottoscharmer.com/sites/default/files/2002_Illuminating_the_Blind_Spot.pdf.

Senge, Peter M., Claus Otto Scharmer, Joseph Jaworski, and Betty Sue Flowers. *Presence: Human Purpose and the Field of the Future.* New York: Doubleday, 2008.

Skipper, Clay. "The Secret to Being a Productive Human: Take More Breaks (and Naps!)," *GQ,* February 8, 2018. www.gq.com/story/the-secret-to-being-a-productive-human-take-more-breaks-and-naps.

Surprise, Kirby. *Synchronicity: The Art of Coincidence, Choice, and Unlocking Your Mind.* Pompton Plains: Career Press, 2012.

Taleb, Nassim N. *Fooled by Randomness: The Hidden Role of Chance in Life and in the Markets.* New York: Random House, 2004.

Tzu, Lao. *Tao Te Ching.* Translated by Stephen Mitchell. New York: Harper Perennial, 1988.

Van Ness, Daniel W., and Karen H. Strong. *Restoring Justice—An Introduction to Restorative Justice.* 4th ed. New Province: Matthew Bender & Co., 2010.

Volk, Anthony A., and Larissa Lagzdins. "Bullying and Victimization among Adolescent Girl Athletes." *Journal of Athletic Training* 11 (2009): 13–31.

Wann, Daniel L. "Examining the Potential Causal Relationship between Sport Team Identification and Psychological Well-Being." *Journal of Sport Behavior* 29, no. 1 (2006): 79–95.

Wann, Daniel L., Paula J. Waddill, Matthew Brasher, and Sagan Ladd. "Examining Sport Team Identification, Social Connections, and Social Well-Being among High School Students." *Journal of Amateur Sport* 1, no. 2 (2015): 27–50.

Wells, Anne J. "Self-Esteem and Optimal Experience." In *Optimal Experience: Psychological Studies of Flow in Consciousness,* edited by Mihaly Csikszentmihalyi and Isabella Selega Csikszentmihalyi, 327–41. New York: Cambridge University Press, 1988.

Wheeler, John A. "Information, Physics, Quantum: The Search for Links." In *Complexity, Entropy, and the Physics of Information,* edited by W. H. Zurek, 309–36. Redwood City: Addison-Wesley, 1990.

Wiseman, Richard. "The Luck Factor." *Skeptical Inquirer,* May 2003. richardwiseman.com/resources/The_Luck_Factor.pdf.

Zukav, Gary. *The Seat of the Soul.* New York: Simon & Schuster, 1989.

Zurek, Wojciech Hubert. "Decoherence, Einselection, and the Quantum Origins of the Classical." *Rev. Mod. Phys.* 75, no. 3 (2003): 715–75.

Zwicky, Arnold. 2005. "More Illusions." *Language Log* (blog), August 17, 2005. itre.cis.upenn.edu/~myl/languagelog/archives/002407.html.

INDEX

A

abundance
  of opportunities or blessings, 152–154
  of universe, 154–156
acausal connecting principle (Jung), 233–236
accountability, in self-managed organizations, 82–83
action
  acting with sense of purpose, 14–17
  aligning with intention, 35–36
  being in flow leads to possibility of, 10
  broadcasting feelings to the world via, 59
  builds toward type of experience, 49
  facilitating unexpected coincidences, 18–19
  in LORRAX cycle, 65
  meaningful actions in symbolic momentum, 99–100
  meaningfulness of, 13–14
  moving us to anticipated qualitative experience, 52
  qualitative experiences from, 33–35
  responsiveness of universe to, 11–14, 23
  synchronicity as, 18–21
  synchronous events as, 7
  that aligns with greater good is resonant with tree of possibility, 135–137
advice process, in self-managed organizations, 69–71
alignment
  action with intention, 35–36
  actions that align with greater good are resonant with tree of possibility, 135–137
  meaningful alignment, 268
  qualitative experience aligning us to outcomes, 44–45
  with synchronicity, 67
  with things as they are, 88
altruism, 178
"amber" stage of development (Laloux), 78
amplitude
  assigning value to possibilities, 216–220
  defined, 267
  describing possibilities of light, 203
  describing properties of objects, 196
anticipated qualitative experience
  actions, thoughts, and feelings moving us to, 52
  defined, 267
  as life urge or intention, 55, 129
  more than intention, 227–229
  prayer and, 127
  selecting, 222–227
  taking the leap and, 123
  tree of possibilities and, 43
  "a world in which environment doesn't matter," 114–115
"apophenia"
  cognitive theory viewing apophenia patterns as illusory, 249
  not explained yet by physical science, 248
  tendency to see meaningful connections between unrelated things, 247–248
archetypes (Jung), 37, 198
Armenta, Christina
  on awe, 154
  on negative feelings as motivators, 157–158
artists, psychological components of flow in, 23

assertiveness, balancing with receptivity, 149–150
athletes, psychological components of flow in, 23
attachment/nonattachment
  flow involving letting go, 67–68
  to known relationships (pros/cons), 78
  nonattachment as quality of flow, 74
authentic self
  finding, 164–169
  making space for what you want, 169–173
authenticity
  allowing by feeling our emotions, 29
  being a part of something bigger than ourselves, 175–179
  creating change and impacting world, 9
  finding your authentic self, 164–169
  freedom to create, 173–174
  making space for what you want, 169–173
  overview of, 163–164
  with people we do not know, 77–78
  in public, 179–185
  relating to others, 46
awe, 154

B

Bareilles, Sara, 181
belief, treating as a muscle, 124–126
The Bell Jar (Plath), 154
Bellows, Alan, 254–255
Berger, Jennifer
  "safe-to-fail experiments," 85
  tools for dealing with complex work environments, 82
biochemistry, synchronicity of, 39
black people
  example of living from heart, 143–145
  prejudice in American society, 145
Bohm, David
  "implicate order" of cosmos ("one-in-all"ness), 202, 204–206
  summarizing quantum mechanics, 194
Bohr, Niels, 199
boldness
  example of getting into graduate program, 109–112

lighting a spark, 138–139
response of universe to our collective choices, 114–115
separating wheat from chaff, 109
shifting probabilities, 102–103
stepping out of comfort zone, 139–140
taking the leap, 121–122
tree of possibilities and, 105–108
uniqueness to the moment, 108
why it is necessary, 104
bottom line, clarity regarding, 68
breaks (pauses), taking breaks for meaningful thinking, 171–173
Buddhist worldview, 127–128
bullying, 78
Burning Man, 142
business contacts, 61–62
business decisions, 86–87
Butler, Margaret, 173–174
"bystander effect," 181

C

Cade, John, 57
careers, Western society focus on, 176
causation, acausal connecting principle (Jung), 233–236
challenge levels, finding flow state and, 24
challenges (global). *see* global challenges
challenges (personal)
  confronting appropriate challenges, 90–91
  flow as balance between challenge and skill, 108–109
  matching challenge and skill, 90–92
  setting and meeting, 128–129
Chalmers, David
  imagining the quality of an experience, 53
  on qualia, 31, 50
  on qualitative experience as building block of nature, 261
change
  being a maker of change (a spark), 239–242
  interactional, 190–191
  "unitary," 190
checking things out, 173–175
choice (personal)

Index

coincidental clues for problem solving, 48
  connecting global challenges to, 8–9
  events in life influenced by, 9
  shaping your world, 45–49
  universe responding to, 9
  world unfolding based on, 47
  your importance/empowerment in the world, 48–49
classical science, comparing with quantum science, 192–195
climate change
  process of changing factors in, 118–120
  symbolic momentum of today's crises, 113, 115–116
cognitive biases, frequency illusion and, 255–256
cognitive theory, viewing apophenia patterns as illusions, 249
coincidence. *see* synchronicity
collective choice
  aligned with individualism, 245
  responsiveness of universe to, 114–115
  role in today's crises, 113–115
collective unconscious (Jung), 36–37
Combs, Allen, 170–171
comfort zone, stepping out of, 139–140
communication style and habits, implementing synchronicity and flow in organizations, 83–84
conscious intention. *see* intention
consistency, of properties, 267
consistent/decoherent histories quantum mechanics, 267
control
  being free of need for, 82
  dropping away in flow, 7
  sharing feelings vs. struggling to, 29
  uncontrollable force of hidden feelings, 56
  as yang quality, 116
core consciousness, 50
core values, 68
correlated superposition of possibilities. *see* quantum correlation
cosmos. *see* universe
counterfactual indefiniteness
  defined, 268
  tree of possibilities and, 42

courage, heart as source of, 140–142
creativity, freedom to create, 173–175
Crucial Conversations, 84–86
Csikszentmihalyi, Mihaly
  on chronic dissatisfaction, 156
  on finding flow state, 23–24
  on flow, 5–7
  on flow as balance between challenge and skill, 108–109
  on flow states, 22
  matching challenge and skill, 90–92
  on nature of work in modern life, 173–174
  on quality of life, 176

D

Damasio, Antonio, 50, 55
dark times, faith during, 117–118
decision making
  business decisions, 86–87
  feelings in, 50
  in self-managed organizations, 87–88
decoherence theory
  defined, 268
  everything acts as an observer, 259
defensiveness, self-inspection without, 152
Dennett, Daniel, 261
desire
  becoming aware of authentic, 166–167
  making space for what you, 169–173
Diener, Ed, 176
disappointment, gratitude and, 159–161
Drucker, Karen, 56
dual-mother archetype, 37

E

education
  uncovering layers of reactive programming, 168
  Western society influencing individuals to focus on own lives, 176
ego
  analogy of ego as a pair of glasses filtering experience, 140–142
  finding your authentic self, 164–169
  habitual fears as egoic habits, 164–165

Einstein, Albert
  special theory of relativity, 201–202
  WWII separates quantum physicists, 199–200
Ekman, Paul, 49–50
emotion. *see also* feelings
  accurately feeling, 46
  amplifying intensity of feelings, 55
  categorizing human (Ekman), 49–50
  comparing with feelings, 50
  falling hostage to hidden feelings, 56
  learning how to feel, 28–30
  as physiological response to our experience, 50
  reactions to experiences, 50–53
  world responding to, 58
employment, changing perspective on, 87
empowerment
  as advocate for change, 184–185
  synchronicity manifesto, 245–246
  your importance in the world, 48–49
entanglement. *see* quantum correlation
entitlement, gratitude and, 158–159
events
  flow influences event outside of ourselves, 23
  retroactive event determination, 207–208
  salient, 264
  singular, 12, 216
existence, properties of things in quantum mechanics, 192–194
experience
  actions in building, 49
  anticipated qualitative. *see* anticipated qualitative experience
  in branching tree of possibilities, 41–45
  emotions as physiological response to, 50
  feelings driving, 53–54
  having or seeking, 50–53
  of meaningful coincidences, 10
  objects as symbols of, 39–40
  possible, 42
  properties and, 32–33, 38, 222
  qualitative. *see* qualitative experience
  reflects our choices, 14
  requires an observer, 190–192
  worldview on impacting collective, 45–49

F

faith
  confidence in the unseen, 126–127
  not limited to religion, 128
  prayer and, 127
  spanning gap between religious and scientific faith, 129
  willingness to step in flow without having all the answers, 129–130
family
  finding balance in family life, 74–75
  importance of, 79–80
fear
  closes us down from possibility, 124
  daunting admonition to conquer, 147–148
  habitual fears as ego habits, 164–166
  heart as source of courage, 140–142
  taking the leap, 121–123
feelings. *see also* emotion
  being driven by, 52
  comparing with emotions, 50
  driving events, 55
  driving synchronicity, 49–55
  as enemy, 58
  hidden, 55–59
  interpreting emotions as, 50–53
  learning how to feel, 28–30
  life orchestrated by, 59
  need to accurately feel, 46
  perceiving emotions through, 50
  playing music with feeling, 144
  pulling us toward meaningful events, 49
  in reasoning and decision making, 50
  as source of synchronicity, 58
femininity
  changing cultural tendencies, 119–120
  polarities of contemporary culture and, 117
Feynman, Richard, 127–128
Fleming, Alexander, 6
flow
  author's concept of, 7
  being in the flow not just going with, 66–67
  defined, 5–6
  dynamic balance between will and surrender, 71

Index

experiencing meaningful coincidences, 10
getting into state of, 8
incorporating synchronicity and, 7–8
interrelationship of synchronicity and, 20–21
as meaningful events or circumstances, 6
physics underlying, 196–198
research on, 21–25
role of values and experiences in, 4
sense of mutual relationship in, 6–7
taming fear of unknown, 9–10
flow channels, 24
Fluke (Mazur), 264
focus, flow and, 21–24
freedom of speech, balancing with standards of decency, 180
freedom to create, 173–174
frequency illusion
  meaningful coincidences and, 248
  synchronicity and, 253–256
Freudian sexual desires, authenticity not descent into repressed desires, 168
friends, finding friends with flow, 76–80

G

Gaertner, Stephen, 3–5
Galileo, 192
gaming, optimistic synchronization in, 208–211
global challenges
  addressing by being authentic, 8–9
  shaping our world through choices, 45–49
  symbolic momentum of today's crises, 113–117
goals. *see* intention
gratitude
  accepting grief, 157–158
  authenticity and, 163
  creating openness to opportunities, 156
  disappointment and, 159–160
  entitlement and, 158–159
  focusing on what really matters to us, 168
  uncovering what you really want, 160–161

greater good, aligning with, 135–137
grief of loss
  allowing, 29–30
  gratitude growing from, 157
  mixed abundance and, 154–156
group will
  aligning with greater good, 135–137
  destructiveness of mob mentality, 137
growth (personal)
  life about learning and growing, 89
  meaningful history selection and, 12
gun violence, 118–120

H

happiness, wealth and, 176–177
heart, living from
  abundance of opportunities or blessings, 152–154
  alignment with the greater good, 135–138
  cosmos responds to what the heart seeks, 158
  ego as a filter of experience, 140
  example of black congregation, 143–145
  gratitude and, 156–161
  heart as lens for viewing reality, 140–143
  irony of white privilege, 145–147
  perspectives on selflessness, 133–135
  purposeful life and, 133–134
  recovering from missed opportunities, 150–152
  selfless synchronicity, 147–150
Heisenberg, Werner, 198–199
Heroic Imagination Project, 181
hidden feelings
  being victim of circumstance, 59
  causing unfortunate synchronicity, 58
  impact of, 55–59
  unearthing, 59
hierarchical organizations, compared with self-managed, 71
Holland, Mark
  on role of play in uncovering what we want, 170–171
  view of synchronicity via lens of mythological trickster, 71

Holocaust, 3–5
human potential, 23

I

"implicate order" of cosmos (Bohm), 202, 204
individualism, aligned with collectivism, 245
"The Influence of Archetypal Ideas on the Scientific Theories of Kepler" (Pauli), 199
inner knowing, synchronicity as, 20
"inner product," mathematical comparison of two properties, 224–225
intention
 aligning action with, 35–36, 49
 choosing according to desired effects, 59
 comparing with anticipated qualitative experience, 227–229
 expecting and meeting obstacles, 93
 or life urge, 55
 navigating possibilities, 43–44
interactions
 determining properties of things, 198–199
 relational nature of, 206
 types of change and, 190–191
interconnectedness, of all things, 177
intimate relationship, 71–76
Irving, Debby, 145
isolation, emergencies breaking spell of, 180

J

Jackson, Frank, 32
Jaworkski, Joseph
 on creating the future, 244
 on finding direction in life, 6
 on listening and then acting, 62, 243
 on relationship, 72
Johnston, Keith
 "safe-to-fail experiments," 85
 tools for dealing with complex work environments, 82
Journal of Biological Chemistry, 39
judgment, coming from limited perspective, 143–144
Jung, Carl
 acausal connecting principle, 233–236
 archetypes and symbolism, 36–40
 influence on Pauli, 199
 on synchronicity, 7, 198
 on synchronicity being dependent on feelings (emotions), 58

K

Kelvin, Lord, 192
Kotler, Steven, 23
Kuhn, Thomas, 90

L

Laloux, Frederick
 on accountability in self-managed organizations, 82–83
 on "amber" stage of development, 78
 on connectedness in self-managed organization, 244
 on self-managed organizations, 69
leadership
 applying LORRAX process to, 81–82
 Jaworski on, 85
 world need for authenticity of, 168
learning, life about learning and growing, 89
Lent, Jeremy, 78
library of heaven, 236–238
life urge, 55. *see also* intention
lifeshocks (McMaster), 65–66
light
 properties of, 201
 scientific research on, 201–202
 speed of, 202–203
 timelessness of, 204–206
listening aspect, of LORRAX process
 finding/determining relevant (meaningful) possibilities, 216
 overview of, 63–64
lithium, discovering calming effect of, 57
LORRAX (Listen, Open, Reflect, Release, Act, Repeat)
 acting phase of, 65
 applying to insecurity, 77
 applying to professional opportunity, 81–82
 balancing assertiveness and receptivity, 149–150

Index

being a maker of change (a spark), 242–243
finding family and relationship balance, 72–75
listening, 63–64
opening, 64
optimizing the noticing of synchronicity, 62–63
reflecting (and releasing), 64
repeating the cycle, 65–66
spacious pause allowing cycle to unfold, 88–89
loss. *see* grief of loss
luck, Wiseman on, 124, 249–253
lymphoma, mustard gas leading to treatment for, 31

M

magical thinking, viewing luck as rational rather than magical, 252–253
Magnus, Albertus, 58
marriage, finding family balance, 74–75
masculinity
 changing cultural tendencies, 119–120
 polarities of contemporary culture and, 117
mass, light lacking, 202–203
materialism, authenticity and, 179
matrix, tool for understanding how objects change, 191
Maxwell, James Clerk
 classical science and, 192
 unifying fields of electricity and magnetism, 201
Mazur, Joseph, 264
McMaster, Ann, 65
meaning
 being the source of, 13
 of events or circumstances in flow, 6
 giving, 11
 hidden emotions and, 58
 of life, 10, 14
 as personal value, 30
 physical world secondary to, 38
 subjective vs. objective, 31–36
 symbolism and, 36–37
 synchronicity and, 4–5

meaningful alignment, 268
meaningful branching, 268
meaningful coincidence. *see also* synchronicity
 objective and subjective meaning, 30–36
 synchronicity as, 18–19
 treating each moment as precious to find, 24–25
meaningful history selection
 "apophenia" and, 247–248
 basic model, 40–45
 building symbolic momentum, 97–98, 112
 defining what you want, 43, 100–102
 determining what is meaningful, 20
 envisioning experiential options, 237–238
 experience reflecting our choices, 14
 feelings driving experiences, 53–54
 growth opportunities in life and, 12
 increasing branches of tree of possibilities, 104
 individual perspective of, 20
 influencing events by life choices, 9, 44, 53–54
 initial action in, 109
 as mirror of our beliefs, 122
 model of, 40–42
 mortgage crisis example, 93
 our role as cocreators of life, 232
 overview of, 213–214
 qualitative experience aligning us to outcomes, 44–45
 responsiveness to collective choices, 114
 responsiveness to underlying feelings, 46
 role of feelings in, 46, 49
 role of obstacles in growth, 92
 selecting anticipated qualitative experience, 222–228
 self-fulfilling nature of good luck, 251
 setting and meeting challenges, 128–129
 spanning gap between religious and scientific faith, 129
 synchronicity and, 233
 tree of possibilities and, 214–221

measurement
  dependent on measured object and one doing the measuring (quantum mechanics), 191–192
  issues in quantum mechanics, 257–258
  scientific, 195–199
Mermin, N. David, 200
Merry, Philip, 86
#metoo movement, example of momentum reaching tipping point, 95–96
mirror, responsive cosmos as, 153–154
Mithraic religion, 36
mixed blessings
  abundance of blessings, 152–154
  grief of loss and, 154–156
  seeing events as, 156
mob mentality, 137
motivation, negative feelings as motivators, 157–158
Murphy's Law, 213, 234
mustard gas, leads to treatment for lymphoma, 31

N

Nazi Germany, 3–5
negative emotions, as motivators, 157–158
Newton, Sir Isaac
  classical science, 192
  research on light, 201
nonexistence, properties of things in quantum mechanics, 192–194
Nonviolent Communication (Rosenberg), 84

O

objective meaning
  defined, 31
  illusory nature of objective worldview, 258–259
  qualitative experience and, 33–35
  quantum mechanics and, 32–33
objective-definite event, 268–269
objects
  properties of. *see* properties
  quantum mechanics on, 32–33
  quantum object, 269
  states suggested by tree of possibilities, 257–258
  as symbols of experience, 38–39
observation
  everything acts as an observer (decoherence theory), 259
  experience requires an observer, 190–192
  relational nature of all interactions, 206
  time viewed as determined (observed) or undetermined (unobserved), 207–208
  what has not been observed is unknown (quantum mechanics), 194–195
  what the world is doing when we are not looking, 258–259
obstacles. *see also* challenges (personal); problems
  expecting and meeting, 93
  role in growth, 92
"one-in-all"ness (Bohm). *see* "implicate order" of cosmos (Bohm)
opening, in LORRAX cycle, 64
openness, with people we do not know, 77–78
opportunity
  applying LORRAX process to, 81–82
  recovering from missed, 150–152
optimistic synchronization
  in gaming, 208–209
  Time Warp Operating System (TWOS) and, 209–211
organizations
  accountability in self-managed, 82–83
  applying LORRAX process to opportunity, 81–82
  changing perspective on employment, 87
  communication style and habits, 83–84
  control in flow and, 82
  methods that rely on synchronicity and flow, 84–86
  nature of decisions made in flow, 87–88
  self-managed vs. hierarchical, 69–71
  tools for dealing with complex work environments (Berger and Johnston), 82
outcomes, being a maker of change (a spark), 239–240

P

parenting, finding family balance, 72–74
Parks, Rosa, 118
past, present, and future, (classical scientific view), 207–208
The Patterning Instinct (Lent), 78
patterns
  physicists trained to look for, 11
  research explaining occurrence of meaningful coincidence, 248
  viewing apophenia patterns as illusions, 249
Pauli, Wolfgang, 199
Penicillium rubens, 6
performers, psychological components of flow in, 23
personal choice. *see* choice (personal)
perspective (point of view)
  God's eye view not valid in terms of modern physics, 206
  on light travel, 203
  relational worldview and, 195, 230
The Phantom Tollbooth, 24
physical world, quantum mechanics in understanding, 192
physics
  development of quantum mechanics, 190
  physicists trained to look for patterns, 11
  of synchronicity, 230–233
  underlying synchronicity, 196–198
  validity of God's eye view and, 206
Pink, Daniel, 172
Plath, Sylvia, 154–155
play, uncovering what we want (Combs and Holland), 170–171
point of view. *see* perspective (point of view)
polarities
  balancing, 115
  quality of contemporary culture and, 115–117
political gridlock
  possibility of immediate change, 119
  symbolic momentum of today's crises, 113–114
position of influence, usefulness of, 164

possibilities. *see also* potential
  of light, 205
  meaningful, 214
  superposition of, 206
  symbolic experience, 237–238
  totality of (equating with heaven), 237
  tree of. *see* Tree of Possibilities
potential
  "amplitude" in describing probability of occurrence, 196
  properties of things in quantum mechanics, 192–194
  quantum mechanics addresses, 192–193
prayer, 127
probabilities
  "amplitude" in describing, 196
  branches of tree of possibilities representing, 216
  relational worldview and, 230–231
  scientific, 195–199
problems. *see also* challenges (personal); obstacles
  choosing self-esteem rather than ease and enjoyment, 91–93
  confronting appropriate, 90–91
  serving development of greater capacity, 89–92
  spacious pause allowing LORRAX cycle to unfold, 88–89
  synchronicity not solving our problems the way we want them to be solved, 159
professional life, balancing with family life, 74
properties
  "amplitude" in describing, 195–196
  dependent on type of measurement performed, 197–198
  experience of objects as, 32–33, 38
  interactions determining, 198–199
  of light, 201
  mathematical comparison of, 224–225
  qualitative experience and, 222
  quantum mechanics and, 192
proto-self, defined, 50
public
  authenticity in, 179–185
  breaking the barrier of silence, 182–184

feeling empowered to advocate change, 184–185
going beyond self-consciousness, 181–182
power of speaking in public, 179–180
public service, 168
purpose
acting with sense of, 14–17
ancient yogic view of, 11
finding, 16–17
meaning of life and, 14
responsiveness of cosmos and, 13
scientific view of universe, 11
searching for, 11–14
puzzle solving (Kuhn), 90

Q

qualia. *see also* qualitative experience
cannot be captured or shared, 53
as raw, felt experiences of being alive, 53–54
as subjecive qualities, 31–32
qualitative experience
alignment with outcomes, 44–45
anticipated. *see* anticipated qualitative experience
body anticipating, 43–44
building blocks of nature, 261
defined, 269
examples of, 114–115
properties and, 222
thoughts and feelings vs., 50
quantum correlation, 269
quantum foundations, 190, 200–201
quantum mechanics
applicability to everyday experience, 195
change processes and, 190–191
comparing classical science with, 192–195
connection with experience of meaningful coincidence, 248–249
on experience of objects as properties, 32–33, 38–39
foundations of, 190
issues in, 257–258
mathematics of, 41
quality of measurement, 191–192
relational, 269
"two-state vector formalism," 213–214
what the world is doing when we are not looking, 258–259
quantum object, 269

R

race
example of black congregation, 143–145
irony of white privilege, 145–147
social identification and, 78–80
white privilege, 142–143
receptivity
balancing assertiveness and, 149–150
LORRAX process fosters, 63
to new relationships, 76
reductionist view, of time, 265
reflecting (and releasing), in LORRAX cycle, 64, 93
Reiher, Peter, 209–210
Reinventing Organizations (Laloux), 69
relational quantum mechanics, 269
relational/relationality
defined, 269
nature of all interactions, 206
perspective-driven view, 195
worldview, 230
worldview advocated by author, 259–261
relationship
example of optimistic synchronization in gaming, 208–209
finding friends with flow, 76–80
flow as sense of mutual, 6–7
healthy, 48–49, 71–76
nature of all interactions, 206
navigating variety of, 79
relaxation, 69
religions
synchronicity influencing, 247
worldviews, 127–128
repeat, in LORRAX cycle, 65–66
research
explaining meaningful coincidence, 248
on flow states, 21–25
on light, 201–202

Index

responsiveness
  cultivating capacity for, 66
  experiences that reflect our deep values, 159
  interpreting responsiveness of cosmos, 232–233
  physics underlying, 196–198
  of universe, 10–14
restorative justice, 150–152
retroactive event determination
  optimistic synchronization, 208–210
  real world described by, 210–211
  time as determined (observed) or undetermined (unobserved), 207–208
rigidity, middle path between rigidity and spontaneity, 68
Roosevelt, Franklin, 200
Rosenberg, Marshall, 84
Rumi, Jalaluddin, 127–128

S

sacrifice, 69
"safe-to-fail experiments," 85
safety/self-preservation, balancing with risk, 71–72
salient (significant) events, in calculating odds of synchronicity, 264–265
Scharmer, C. Otto, 85
Schrödinger's cat, 258, 260
science
  "apophenia" and, 248
  classical vs. quantum, 192–195
  experience requires an observer, 190–192
  foundations of synchronicity and flow, 189
  prayer and faith and, 127–128
  probabilities and measurement, 195–199
  "shut up and calculate" paradigm, 199–201
  timelessness of light, 201–206
  view of purpose of universe, 11
  worldviews, 127–128
  yang quality of technological civilization, 115

scientific research. *see* research
The Seat of the Soul (Zukav), 35–36
self-consciousness, going beyond in public setting, 181–182
self-esteem, choosing over ease and enjoyment, 91–93
self-importance/self-promotion, being part of something bigger than ourselves, 175–179
self-knowledge/self-awareness, 163–164
selfless synchronicity, 147–150
selflessness
  focus on others as means of overcoming fear, 147–149
  perspectives on, 133–135
self-managed organizations
  accountability in, 82–83
  compared with hierarchical, 71
  connectedness in, 244
  decision making in, 87–88
Senge, Peter M., 84
separateness, 140
September 11, 2001, 28
serendipity, 269
sexuality, authenticity and, 168
shared purpose, 182–183
"shut up and calculate" paradigm (Mermin), 199–201
silence, breaking the barriers in public, 182–184
Simple Habits for Complex Times (Berger and Johnston), 82
singular events
  as forks in the road, 12
  in tree of possibilities, 216
singular moment, 270
skills
  flow as balance between challenge and, 108–109
  matching challenges to, 24, 90–92
  survival skill, 78
social identification, 78
social issues, symbolic momentum of today's crises, 115–116
social sciences, on synchronicity, 247–248
Source: The Inner Path of Knowledge Creation (Jaworski), 85

space, special theory of relativity, 202
spacious pause, stopping to feel situation, 88
spark/catalyst, being a change maker
   example (author's musical participation), 239–242
   LORRAX process and, 242–243
spiritual teaching, finding essence of, 13
spontaneity, middle path between rigidity and, 68
sports psychology, 78
stakeholders, in self-managed organizations, 70
statistics, calculating odds of synchronicity, 263–266
Steindl-Rast, David, 127–128
stress
   being in the flow and, 69
   parenting and, 72–74
subjective meaning
   personal interpretation and, 31
   qualia and, 31–32
   qualitative experiences and, 32–33
superposition
   defined, 270
   measurement of possible states, 194
   of possibilities, 206
   reasons why it applies to everything, 204
   timelessness of light and, 204–205
surrender, as yin quality, 116
survival skills, 78
symbolic momentum
   boldness shifting probabilities, 102–103
   change coming from inside ourselves, 119–120
   of contemporary crises, 113–117
   courageous leaps, 121–123
   of destructive cycles, 101–102
   example (#metoo movement), 95–96
   example (getting into graduate program), 109–112
   example (immediate change), 118
   example (job search), 98–99
   faith in flow, 126–131
   meaningful actions in, 99–100
   overnight change, 117–118
   response of universe to our collective choices, 114–115
   separating wheat from chaff with bold action, 109
   shifting probabilities, 102–103
   treating belief as a muscle, 124–126
   tree of possibilities and, 105–108
   unfolding of momentum, 96–98
   uniqueness of boldness to the moment, 108
   why boldness is necessary, 104
symbolism
   of experiences, 37–38
   Jungian archetypes, 36–37
   objects as symbols of experiences, 39–40
   possibilities of symbolic experience, 237–238
   qualitative nature of symbolic events, 97–98
   symbolic action in meaningful history selection, 226
   of tree of possibilities, 225
   underlying quantum mechanics and synchronicity, 199
synchronicity
   acausal connecting principle (Jung), 233–236
   aligning with, 67
   of biochemistry, 39
   calculating odds of, 263–266
   creating situations of, 242
   debunking, 247–256
   defined, 229–230, 270
   described by Jung, 7
   feelings driving, 49–55, 58
   flow and, 7–8, 20–21
   frequency illusion and, 253–256
   hidden feelings causing unfortunate, 58
   leads to flow, 25
   luck and, 249–253
   as meaningful coincidence, 7
   meaningful events and, 4–5, 17–18, 230
   meaningful history selection and, 233
   negative consequences, 231–232
   neutral quality of, 234, 236
   noticing/paying attention to, 4, 62–63
   Pauli's interest in, 199
   personal nature of, 246
   physics of, 196–198, 230–233
   quantum mechanics and, 248–249
   as reflection or mirror, 29

research on meaningful coincidence, 248
scientific view of, 4
social sciences on, 247–248
synchronicity manifesto, 245–246
tree of possibilities and, 220–221, 228
world shaped by, 18–19
"Synchronicity and Leadership" (Merry), 86
Synchronicity: The Inner Path of Leadership (Jaworski), 85
systemic oppression, of black people, 145

T

Tao Te Ching, 125–126
Taoist worldview, 127–128
team identification, in sports psychology, 78
technological civilization, yang quality of, 115
theory of meaningful history selection. *see* meaningful history selection
thoughts, triggered by emotions, 50–53
time
　as collection of salient moments, 265
　determined (observed) or undetermined (unobserved), 207–208
　special theory of relativity, 202
　taking breaks for meaningful thinking, 171–173
　timelessness of light, 204–206
　timelessness of tree of possibilities, 225–226
Time Warp Operating System (TWOS), 209–211
timelessness of light
　relational worldview and, 260
　retroactive event determination, 207–208
　superposition and, 204–206
　"two-state vector formalism," 213–214
timing, flow and, 21
"trait self-control," 171–172
tree of possibilities
　ability to feel (anticipate) a potential experience impacts, 223–224
　abundance afforded by, 154–155
　assigning value (amplitude) to possibilities, 216–220
　boldness shifting probabilities, 105–107

branching nature of, 214–216
defined, 270
emotion and, 55
greater good resonant with, 135–137
library metaphor, 236–238
mathematical comparison of two properties, 224–225
meaningful history selection and, 43–44
object states and, 257–258
overview of, 41–42
representing meaningful outcomes, 42–43
soul's journey along (Zukav), 235
synchronicity and, 220–221, 228
timelessness of, 225–226
trust
　living in the flow, 246
　people we do not know, 77–78
　speaking up in public and, 182–183
　whatever path you choose, 9–10

U

unconscious feelings, impact of, 55–59
undetermined states, 270
"unitary" change, 190
universe
　friendliness of, 11–12
　living in a participatory, 59
　relationship as organizing principle (Jaworski), 72
　responds to collective choices, 114–115
　responds to personal choices, 9
　responsive cosmos as mirror, 153–154
　responsiveness in building our lives, 46
　responsiveness of, 10–14, 23, 66
unknown, what has not been observed (quantum mechanics), 194–195
U-process (Senge), 84
U-shaped innovation process (Scharmer), 85

V

viewpoint. *see* perspective (point of view)
violence
　inner nature of process of change, 118–120
　symbolic momentum of today's crises, 113–114

W

Waking Up White (Irving), 145
wave function, 270
wealth, happiness and, 176–177
Western society, 176
Wheeler's delayed choice experiment, 207
When: The Scientific Secrets of Perfect Timing (Pink), 172
white privilege
  author's upbringing and, 142–143
  irony of, 145–147
winners, everyone wins in flow consciousness, 242–245
Wiseman, Richard, 124, 249–253
wonder, mixed abundance and, 154
work
  dealing with complex work environments (Berger and Johnston), 82
  nature of work in modern life, 173–174
worldview
  hidden emotions shaping, 58
  illusory nature of objective worldview, 258–259
  impact on collective experience, 45–49
  quantum mechanics and, 257
  relational, 230, 259–261
  religious and scientific traditions, 127–128
  role of faith in, 130–131

Y

yin-yang polarity, 115–117
yogic sciences, 11, 127–128

Z

Zukav, Gary
  on challenge of dealing with negative aspects of self, 166
  on choosing intentions according to desired effects, 59
  on intention, 52
  on soul's journey along branching tree of possibilities, 235

ABOUT THE AUTHOR

SKY NELSON-ISAACS is a physicist, musician, teacher, parent, and activist. From an early age, he was exposed to his parents' spiritual teacher, Sri Swami Satchidananda, whose influence led him to an interest in understanding the purpose of existence. This influence has shaped his interest in physics as a way to understand what life is about.

Nelson-Isaacs earned his BA in physics from the University of California at Berkeley, his teaching credential in physics from Sonoma State University, and his MS in physics from San Francisco State University, with a thesis on string theory. He has many years of experience as a physics and math instructor, and he has also worked in the software industry as a project coordinator and quality control engineer for a financial services company. Nelson-Isaacs's current work involves research in the field of quantum foundations, focusing on the principles discussed in this book. He has published a growing body of peer-reviewed work developing the foundation of a theory of synchronicity.

Nelson-Isaacs has earned recognition as a singer, songwriter, and performer, and he has studied piano and guitar since childhood. He is an in-demand performer among the nationwide New Thought community and in his native San Francisco Bay Area community. His songs tell of the positivity and well-being that comes from delving sincerely into the quest for self-knowledge.

Nelson-Isaacs has become increasingly politically active over the past decade, finding a deep satisfaction (although not exactly enjoyment) in it. He believes that an understanding of synchronicity will reshape the way we think about our personal and professional lives when we realize we can find more harmony and success by seeing the unexpected coincidences that serve as guidance on our path. He is available for speaking engagements and concerts. His greatest love is spending time with his wife, Dana, and his daughter, Ellie.

He can be found online at www.skynelson.com.

About North Atlantic Books

North Atlantic Books (NAB) is an independent, nonprofit publisher committed to a bold exploration of the relationships between mind, body, spirit, and nature. Founded in 1974, NAB aims to nurture a holistic view of the arts, sciences, humanities, and healing. To make a donation or to learn more about our books, authors, events, and newsletter, please visit www.northatlanticbooks.com.

North Atlantic Books is the publishing arm of the Society for the Study of Native Arts and Sciences, a 501(c)(3) nonprofit educational organization that promotes cross-cultural perspectives linking scientific, social, and artistic fields. To learn how you can support us, please visit our website.